The PLL Synthesizer Cookbook

by Harold Kinley

TAB BOOKS Inc.
BLUE RIDGE SUMMIT, PA. 17214

FIRST EDITION

FIRST PRINTING—SEPTEMBER 1980
SECOND PRINTING—APRIL 1981

Printed in the United States of America

Library of Congress Cataloging in Publication Data

Kinley, Harold
The PLL synthesizer cookbook.

Includes index.
1. Citizens band radio—Equipment and supplies.
2. Phase-locked loops. I. Title.
TK6570.C5K56 621.3846 80-19966
ISBN 0-8306-9707-1
ISBN 0-8306-1243-2 pbk.

Preface

The first time I was faced with repairing a phased-lock loop (PLL) circuit in a radio I realized just how little I knew about PLL synthesizers. With a little luck and a lot of time, I managed to get the thing repaired though. But we technicians and hobbyists cannot always depend on luck and we certainly cannot afford to spend lots of time on one repair job. Besides, I had always prided myself on using logical troubleshooting procedures based on an understanding of the particular circuit at hand. Troubleshooting by the "hit or miss" method never suited my style.

While I was involved in my first PLL encounter, I decided then and there that I would do everything possible to learn all I could about these new PLL synthesizers. After much searching for books on the subject, I was disappointed at finding practically nothing written on the subject. So I turned to the manufacturers of the PLL ICs for data sheets, application notes or whatever information they could supply on these IC products. I spent quite some time pouring over this technical data. I also experimented on PLL circuits for that "hands on" experience. With all this researching, I finally reached a point where I was comfortable with PLL synthesizers. I no longer had to regard the PLL circuit as a *black box*. I could now apply my logical troubleshooting method to these circuits too.

The thought then occurred to me: There must be thousands of technicians like myself who want to learn how PLL synthesizers work. I then decided to put the results of my research of PLL synthesizers into a book.

The first chapter is an introduction to the basic PLL synthesizer circuit. The rest of the book covers specific PLL ICs and circuits in detail. The appendices give information on still more PLL ICs.

Several semiconductor and CB manufacturers supplied information and material for the book. I therefore wish to express my appreciation to the following companies for their assistance:

Cobra Communications (Dynascan Corp.)—Bruce Diamond
E. F. Johnson Co.—Robert Cervenka
Fairchild Camera & Instrument Co.
Faron/Courier (Resdel Industries)—Miguel Santana
Melco (Mitsubishi Electric Corp.)
Midland International Corp.
National Semiconductor Corp.
NEC (National Electric Co.)
OKI Semiconductor
Panasonic (Matsushita Electronics Corp.)
Pearce—Simpson (Division of Gladding Corp.)—Earle Smith
President Electronics/American Radio—Rex Trobridge
Radio Shack Store, Kingstree, SC (Haigler Hardware)
Toshiba America

Harold Kinley

Contents

1 **Introduction to PLL Synthesizers** 7
Programmable Dividers—Fixed Dividers—Phase Detectors/Lock Detectors—Charge Pumps and Filters—Voltage-Controlled Oscillators—Crystal-Controlled Oscillators—Mixers—The PLL as a System—Some Practical Considerations

2 **The μPD2816C Digital Synthesizer** 24
Special Features—Internal Structure—A Typical Application—Cobra 1000 GTL CB Synthesizer

3 **The SC42502P/3001-201 Digital Synthesizer** 42
Special Features—Internal Structure—A Typical Application—E.F. Johnson Messenger 4140 Synthesizer

4 **The REC86345 Digital Synthesizer IC** 63
Special Features—Internal Structure—Fanfare 190 DF PLL Synthesizer

5 **The PLL02A Digital Synthesizer IC** 79
Special Features—Internal Structure—Practical Application—Midland 77-380 CB PLL Synthesizer

6 **The MM55104N PLL IC** 101
Reference Oscillator—Reference Divider—Programmable Divider—Phase Detector—Practical Application—Pearce-Simpson Super Tiger 40A PLL Synthesizer

7 **The μPD858C PLL IC** 123
Special Features—Internal Structure—Practical Application—President Model Madison

8 **The μPD861C PLL IC** ... 157
Special Features—Internal Structure—Practical Application—Realistic TRC-424

9 **The MM55106 PLL IC** ... 178
Special Features—Internal Structure—Practical Application—The Regency CB-501 PLL Circuit

10 **The MN6040 PLL IC Series** ... 201
Special Features—Internal Structure—Robyn SX-402D PLL Synthesizer

11 **Toshiba PLL ICs** ... 218
TC5080P—TC5081P—TC5082P—The Tram D12 PLL Synthesizer—Sharp Model CB-2260 PLL Synthesizer

Appendices

A **The PLL03A PLL IC** ... 240
Pin Functions—Application

B **The M58473P PLL IC** ... 243
Pin Functions—Application

C **The MSM5807 PLL IC** ... 247
Pin Functions—Application

D **National Semiconductor PLL ICs** ... 249
MM55104, MM55106, MM55107, MM55114, MM55116 PLL Frequency Synthesizer—MM55108, MM55110 PLL Frequency Synthesizer with Receive/Transmit Mode—MM55109, MM55111 PLL Frequency Synthesizer and Channel Programmer—MM55122 Serial Data/PLL Frequency Synthesizer—MM57150 Standard CB Radio Controller

E **Fairchild 11C84 PLL IC** ... 270
Input and Output Functions—Single Crystal System—Dual-Crystal System—Cost Versus Performance Trade-Off—40-Channel CB Synthesizer—Transmit Frequency Generation, Direct Versus Derived

Index ... 278

Chapter 1
Introduction to PLL Synthesizers

In this chapter you'll learn what various stages make up a typical PLL synthesizer. Then after all stages are discussed, a complete PLL synthesizer will be analyzed. The basic stages which make up a PLL synthesizer are:

- ☐ Programmable divider
- ☐ Fixed divider
- ☐ Phase detector
- ☐ Charge pump and filter
- ☐ Voltage-controlled oscillator
- ☐ Crystal-controlled oscillator
- ☐ Mixers

PROGRAMMABLE DIVIDERS

The programmable divider is defined as a frequency divider that can be programmed to divide its input frequency by a certain divisor (N). The actual divisor is determined by the application of proper voltage levels to the various *program* pins on the IC device. There are two voltage levels used on the program pins, the HIGH level (usually near the IC supply voltage) and the LOW level (at or near ground potential). The HIGH level is usually referred to as binary 1 and the LOW level as binary 0. This will be fully explained shortly.

The proper voltage levels are usually applied to the program pins through the channel selector switch directly. Hence, the various positions of the channel selector switch apply HIGH and

LOW levels to the program pins in various combinations. Each different combination or code programs the divider to divide the input frequency by a certain divisor, N. In order to understand how this programming is achieved, you will need a basic understanding of the binary numbering system.

Pure Binary System

The binary number system uses only two different digits to represent any decimal number. The two digits are binary 1 and binary 0. Because only two different digits are used, the system is said to have a base of 2. For comparison, the decimal system uses 10 different digits. It therefore has a base of 10. Both the decimal system and the binary system are weighted systems. This simply means that the relative positions occupied by the digits determine the weight the digit carries. In the decimal system the weights are units, tens, hundreds, thousands, ten thousands, and so on for whole numbers. For example, in decimal number 5,362, the weight of the position occupied by the 2 is units, the weight of the position occupied by 6 is tens, the weight of the position occupied by the 3 is hundreds, and the weight of the position occupied by the 5 is thousands. The weight of each position is 10 times the weight of the position to its right.

The binary system uses only two different digits, 0 and 1, so the weight of each position or bit of a binary number is twice the weight of the bit to its right. For example, in the positions, ..., ..., ..., 256, 128, 64, 32, 16, 8, 4, 2, 1, each bit position has a weight which is twice that of the bit position to its right. The bits increase in weight from the right to the left by a factor of 2. The bit to the far right is called the *least significant bit* (LSB) and the bit to the far left is called the *most significant bit* (MSB). Here is a typical binary expression: 10010111. Unless otherwise noted, assume that the bit to the far right is the least significant (LSB) and the bit to the far left is the most significant (MSB). How do you determine the decimal number represented by this binary expression? Here's how: Simply write down the weights of the positions occupied by the binary 1 and then add them together to find the decimal equivalent. If necessary, write down the weights of the various positions above the binary digits like this: 1 0 0 1 0 1 1 1. Then simply add the weights of the bit positions where the binary 1 appears. In this example, this would be 128 + 16 + 4 + 2 + 1, or 151. So the binary expression 10010111 is the equivalent of decimal number 151. The number of bit positions required to conver from decimal to binary form varies with the value of the decimal number. For instance, the binary equivalent of 2 could be written as simply binary 1 0. The decimal number 9 would be binary 1 0 0 1.

In converting a decimal number to binary, the following procedure will work: Write out the various weights of the bit positions as follows; ..., ..., 256, 128, 64, 32, 16, 8, 4, 2, 1. You need only expand out to a weight position that is equal to or smaller than the number to be represented. For example, to represent decimal number 249, you need only to expand the positions out to 128, because the next position (256) is larger than the decimal quantity to be represented. At this point, there will be eight bits required to represent decimal number 249. Each bit will be represented by either binary 1 or binary 0. All you need to do now is find out what the state of each position will be. Here is the procedure: From 249, subtract the weight of the largest bit position to be used. It has already been determined that the largest bit position is 128, so 249-128=121. Place binary 1 over position 128. Repeating this procedure, the largest bit position that doesn't exceed 121 is 64, so place binary 1 over 64 and subtract 64 from 127. This yields 121-64, or 57. The largest remaining bit position that doesn't exceed 57 is 32, so place binary 1 over 32 and subtract 32 from 57. This yields 57-32, or 25. The largest remaining bit position that doesn't exceed 25 is 16, so place binary 1 over 16 and subtract 16 from 25. This yields 25-16, or 9. The largest remaining bit position that doesn't exceed 9 is 8, so place binary 1 over the 8 and subtract 8 from 9 to yield 9-8, or 1. Place a binary 1 over the 1. Then over the unused positions, place binary 0. The expression would look like this:

1 1 1 1 1 0 0 1
128-64-32-16-8-4-2-1.

In the calculations, we used 128, 64, 32, 16, 8, and 1. All these have binary 1 over them. The only two positions not used are the 4 and 2 positions, so they would have a binary 0 over them to indicate that they are not used. The final binary expression is 11111001. This is the equivalent of 249. This system of binary numbering is called the *pure binary system*. This system is used extensively in programming the programmable dividers used in PLL synthesizers.

Binary Coded Decimal System

The binary coded decimal (BCD) system is a modified form of the pure binary system. This system uses the weighted 8-4-2-1 bits to represent each individual digit of a decimal number. Therefore, four binary bits are required for each decimal digit. For example, representing the decimal number 249 in BCD form would require 3 × 4, or 12 binary bits. The 2 would become 0010; the 4, 0100; and the 9, 1001. This would be written out as: 0010.0100.1001. The dots between each binary group indicates that the BCD code is used. The BCD system requires more bits to represent a given decimal number than does the pure binary system. For example, the number 249 in the pure binary system requires only 8 bits (11111001) whereas 12 bits are required in the BCD system (0010.0100.1001). This factor makes the circuitry for BCD systems more complex, but a real advantage of the BCD system is that it is easier for most of us to make the conversion between decimal numbers and binary numbers in BCD form.

Since there are only 10 different codes used in the BCD system, one who uses the system regularly soon learns to recognize the decimal equivalents at a glance. The conversion codes for the 10 different digits are shown in Table 1-1. This BCD system is also used in some PLL ICs to program the programmable dividers. Some dividers can be programmed with either the BCD or the pure binary system.

Decimal number	BCD Equivalent
0	0000
1	0001
2	0010
3	0011
4	0100
5	0101
6	0110
7	0111
8	1000
9	1001

Programming the Divider

Figure 1-1 is a block diagram of a programmable divider. The pins labeled P0 through P4 are the programming pins of the device. Each pin has a specific *weight*. The individual pin weights are P0=1, P1=2, P2=4, P4=8, and P4=16. The divider will divide the input frequency by the number N, which is programmed into the divider. The number programmed is determined by adding together the weights of the pins at the binary 1 level. (binary 1 = 5 volts, and binary 0 = 0 volts or ground). You can see that P4 is connected directly to the 5-volt supply, so P4 will always be at binary 1 level. This means that its weight of 16 will always be used in determining the divisor N. The other program pins (P3 through P0) are connected to a switch that can either ground the pin or apply 5 volts to it; that is, the switch can apply either binary 1 or binary 0 to the pin. Thus, the device can be programmed by the various combinations of switch positions. Because there are four switches used and each switch has two positions, there are 2^4, or 16, possible combinations. Therefore, 16 different divisors N are possible using this arrangement. Table 1-2 shows the 16 different combinations.

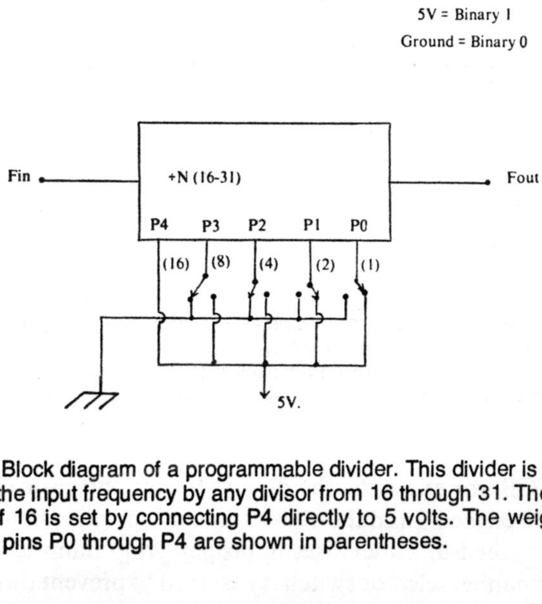

Fig. 1-1. Block diagram of a programmable divider. This divider is capable of dividing the input frequency by any divisor from 16 through 31. The minimum divisor of 16 is set by connecting P4 directly to 5 volts. The weights of the program pins P0 through P4 are shown in parentheses.

Look closely at the switch positions in Fig. 1-1 and see if you can determine the divisor, N, which is programmed into the divider. If necessary, use the chart in Table 1-2. The divisor, N, is determined by simply adding the weights of the pins that are at the binary 1 level. In the example in Fig. 1-1, the pins at binary 1 level are P4 (always at binary 1), P1, and P0. Simply add the weights of these pins as shown below to find the divisor N:

P4= 16
P1= 2
P0= 1

19 (total) = divisor N

The divider will divide its input frequency by 19, so the frequency appearing at the output of the divider will be F (in) ÷ 19 = F (out). This divider is programmed using the pure binary code.

Let's look at a divider that is programmed using the BCD system. An example is shown in Figure 1-2. In this illustration, the divider is shown to be programmed for channel 21. An internal *code converter* is built into the device. This converter in turn determines the proper divisor, N, for each channel. For a

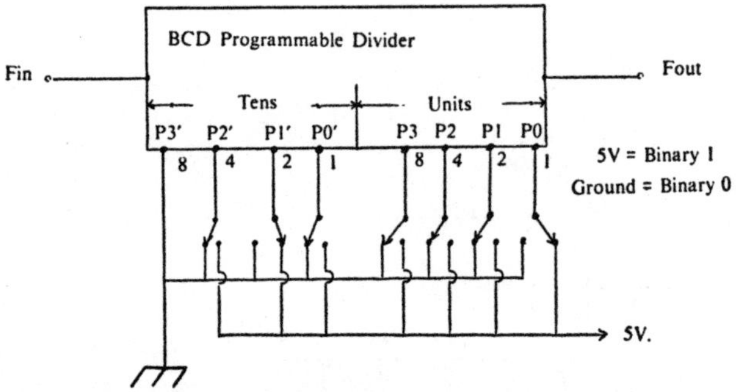

Fig. 1-2. Block diagram of a BCD programmable divider. The tens digit is set by PO' through P3' and the units digit is set by P0 through P3. The divider here is programmed for channel 21. The program lines shown are actually connected to a code converter, which in turn applies the proper binary levels to the internal programmable divider.

40-channel CB, for example, the digit in the tens position cannot be allowed to exceed 4 and the digit in the units position cannot be allowed to exceed 9. Otherwise, improper programming would result. The channel selector switch is designed to prevent this from occurring. Some IC PLL devices incorporate a special circuit to detect when the program pins are improperly programmed. This detector output can be used to kill the transmitter output when *misprogramming* occurs, caused by loose contacts on the channel switch or other reasons. This prevents the radio from radiating improper frequencies.

Referring again to Fig. 1-2, you can see that the tens position has 4 program lines: P0', P1', P2', and P3', with respective weights of 1, 2, 4, and 8. The units position also has 4 program lines: P0, P1, P2, and P3, with respective weights of 1, 2, 4, and 8. The binary level on pins P0 through P3 determines the number in the units position, while the binary level applied to pins P0' through P3' determine the number in the tens position. Table 1-2 shows the allowable combinations for the numbers 1 through 40. Notice that P3' is always at binary 0 so it is connected directly to ground.

Programmable dividers used in CB PLL synthesizers generally aren't capable of dividing input frequencies in excess of 5 MHz. These devices are usually constructed with CMOS circuitry for low power dissipation.

Table 1-2. Different Program Combinations Applied to Divider of Fig. 1-1.

PIN	P4	P3	P2	P1	P0
Weight	16	8	4	2	1
N=16	1	0	0	0	0
N=17	1	0	0	0	1
N=18	1	0	0	1	0
N=19	1	0	0	1	1
N=20	1	0	1	0	0
N=21	1	0	1	0	1
N=22	1	0	1	1	0
N=23	1	0	1	1	1
N=24	1	1	0	0	0
N=25	1	1	0	0	1
N=26	1	1	0	1	0
N=27	1	1	0	1	1
N=28	1	1	1	0	0
N=29	1	1	1	0	1
N=30	1	1	1	1	0
N=31	1	1	1	1	1

FIXED DIVIDERS

The fixed dividers used in PLL circuits usually operate with higher input frequencies than the programmable dividers. The reference oscillator frequency (usually 10.240 MHz) is divided by the fixed divider to yield the basic reference frequency. The basic reference frequency is usually 10 kHz. Sometimes, though, 5 kHz is used as the reference. For a 10 kHz reference frequency, the divider would have to divide the reference oscillator frequency (10.240 MHz) by 1024. Some fixed dividers also have the divide-by-2 output available at a terminal of the IC. This would be 10.240

MHz divided by 2, or 5.12 MHz. This frequency can be multiplied and used in some PLL systems to mix with the voltage-controlled oscillator frequency for down conversion. You will see an example of this later.

PHASE DETECTORS/LOCK DETECTORS

The output of the programmable divider is fed to an input to the phase detector or comparator. The 10 kHz reference signal from the fixed divider is fed to a separate input of the phase detector. The phase and frequency of the two inputs are compared, and a pulsed output is produced. The pulse width and spacing of the output pulses depend upon the relative frequency and phase of the detector inputs. The output of the phase detector is used to control the voltage-controlled oscillator. However, the pulse output of the detector must be properly filtered before being applied to the voltage-controlled oscillator. Usually associated with the phase detector is an output which indicates a *locked* or *unlocked* condition. If an unlocked condition exists, the lock detector output provides a signal for stopping the transmitter output. This prevents signal radiation at improper frequencies.

CHARGE PUMPS AND FILTERS

A typical CMOS charge pump is shown in Fig. 1-3. The function of the charge pump is to feed the pulses from the phase detector to an RC (resistor-capacitor) filter nerwork to filter out the pulsations. Then the smooth DC voltage can be used to control the voltage-controlled oscillator. Notice that the charge pump accepts two inputs from the phase detector and has a single ended output. Some ICs incorporate an *active filter*. This is a specially designed amplifier with a controlled feedback network. This amplifier filter resists any rapid or high-frequency changes in its output. The result is a low-pass filter.

VOLTAGE-CONTROLLED OSCILLATORS

As the name implies, a voltage-controlled oscillator (VCO) is an oscillator whose frequency is governed by a control voltage. The component which makes this possible is the *varactor diode.* All diodes exhibit a certain amount of capacitance when they are reverse biased. The varactor diode is simply a junction diode that is carefully manufactured so that the capacitance varies nearly linearly with a changing reverse bias or control voltage. The

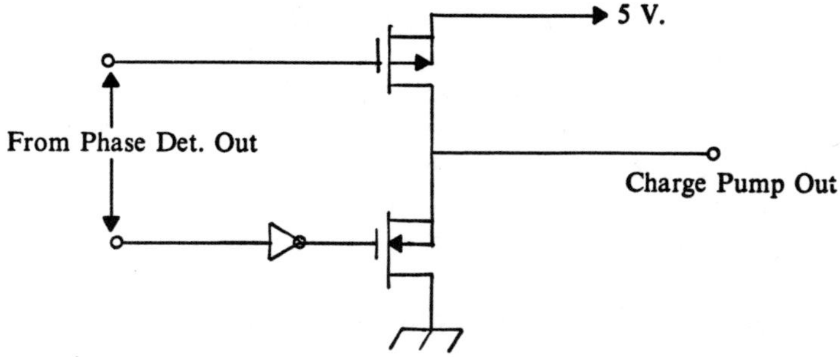

Fig. 1-3. A typical charge pump. Notice the dual inputs from the phase comparator is converted to a single ended output.

schematic of a voltage-controlled oscillator is shown in Fig. 1-4. Diode D1 is the varactor diode. Notice the symbol used to represent the varactor. The capacitor in the circle indicates the diode is a varactor.

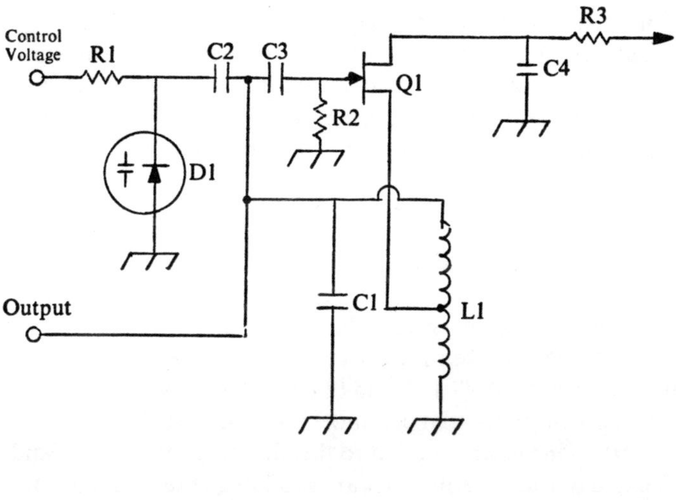

Fig. 1-4. A voltage-controlled oscillator in the Hartley configuration. A control voltage applied to varactor diode D1 through resistor R1 controls the oscillator frequency.

This oscillator is basically a Hartley-type oscillator. The frequency-determining components consist of the varactor, D1; C2; C1; and L1. Because D1 forms part of the frequency-determining network, any change in the capacitance of D1 will change the output frequency of the VCO. The capacitance of D1 can be changed by varying the positive control voltage applied to D1 through resistor R1. The control voltage would have to be positive in order to reverse bias the varactor. If the varactor were connected oppositely a negative control voltage would then be required. Normally, PLL circuits use positive control voltages.

CRYSTAL-CONTROLLED OSCILLATORS

Anyone reading this book should be familiar with crystal-controlled oscillators. One of the main advantages of PLL synthesizers is that the use of crystals is minimized. Some PLL synthesizers require three or more crystals; some require only one. The basic reference frequency is always derived from a crystal-controlled oscillator. The reference oscillator in the majority of PLL synthesizers operates at 10.240 MHz. This usually serves as second oscillator frequency for the receiver, also.

Other crystal-controlled oscillators in PLL synthesizers are usually used to mix the VCO frequency for up or down conversion. Usually the VCO operates at higher frequency than the input of the programmable divider can accept. So the VCO frequency must be converted down to a frequency which can be handled by the divider. These operations are discussed in detail in the following chapters.

MIXERS

Mixer circuits are used in most PLL synthesizers. The operation of these circuits is no great mystery to the average technician. There are, however, some special IC mixers used in PLLs which deserve some special attention. Two commonly used IC mixers used in PLLs are the C3001-A and the TA7310P. Basically, these devices consist of dual-balanced mixers and an oscillator. When two input frequencies are applied to this device, both the sum and the difference of these frequencies are available at separate output terminals. Figure 1-5 shows the basic setup for the C3001-A.

The frequency of the oscillator is controlled by the crystal (X1) connected to pin 1 through capacitor C1. A portion of the oscillator signal is fed to pin 2 through the voltage divider network C2-C3. This signal at pin 2 is identified as F1. Another signal (F2) is fed to pin 4. Now the sum of F1 and F1 is taken from pin 6 and the difference of F1 and F2 is taken from pin 9. These mixers are used to advantage, as you will see in a following chapter.

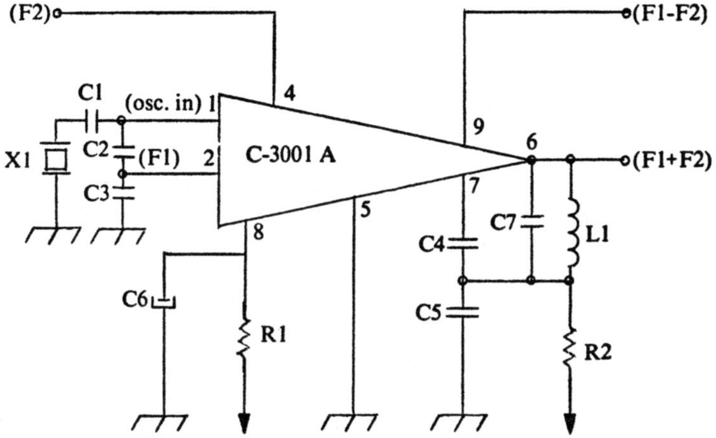

Fig. 1-5. A typical setup for the dual balanced mixer/oscillator IC, C3001-A. Both the sum and the difference of inputs f1 and f2 are available at separate output terminals.

THE PLL AS A SYSTEM

Now that the function of the various individual PLL stages have been discussed, it is time to analyze the PLL system as a whole. During this discussion, refer to the block diagram in Fig. 1-6. the basic reference frequency is derived from the reference oscillator. The 10.240 MHz signal from the reference oscillator is fed to the fixed divider where the frequency is divided by 1024 to yield 10 kHz. This 10-kHz signal is then used as the basic reference frequency. The reference oscillator also serves as the second receive oscillator.

The programmable divider divides its input frequency by divisor N. The N number is determined by the binary levels applied to the program inputs, P0 through P7. The weights of P0-P7 are:

P0 = 1
P1 = 2
P2 = 4
P3 = 8
P4 = 16
P5 = 32
P6 = 64
P7 = 128

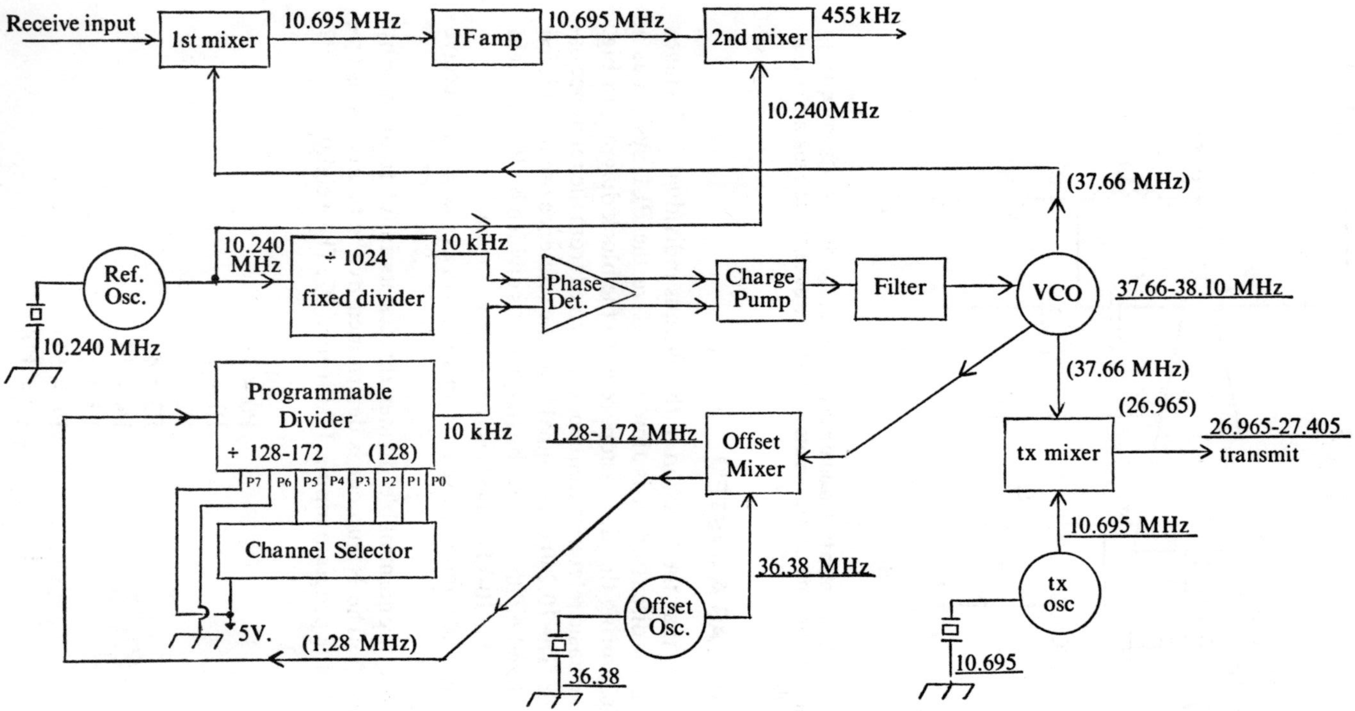

Fig. 1-6. The complete block diagram of a typical PLL synthesizer. The frequencies shown in parentheses are for channel 1 operation.

Notice that P7 does not connect to the channel selector switch. Instead it is connected directly to the 5-volt supply source. This means that P7 (weight of 128) will be at the binary 1 level at all times, regardless of the channel selector position. This automatically sets the *minimum divisor N* at 128. Also notice that P6 is not connected to the channel selector switch. It instead connects directly to ground, thus keeping P6 at the binary 0 level at all times. The weight of P6 (64) will therefore never be used. The other program lines, through P0, P5, all connect to the channel selector switch, so the binary level on these lines will depend upon the channel selector position.

The divider is set up to have an N range of 128 through 172. Actually, with P6 grounded, the divider could possibly be programmed to divide by up to 191 if all the other program lines were set to binary 1. See the calculation:

P0 = 1
P1 = 2
P2 = 4
P3 = 8
P4 = 16
P5 = 32
P7 = 128

191 total N

However, the maximum divisor needed for this particular setup is 172. So the channel selector switch is designed to limit the *maximum divisor N* to 172.

Let's go through the analysis of the complete PLL using channel 1 as an example. The VCO operates in the 37-MHz range. The VCO itself serves directly as the first receive oscillator. The first intermediate frequency (i-f) is 10.695 MHz, so the VCO must operate at a frequency that is equal to the sum of the channel frequency and the 10.695-MHz first i-f. For CB channel 1, for example, the VCO frequency would be 26.965+10.695, or 37.660 MHz. In the transmit mode, the VCO frequency is mixed with a 10.695-MHz signal to yield the transmit frequency. For channel 1 this would be 37.660-10.695, or 26.965 MHz. Notice that this is just the reverse of the mixing operation for the receive mode.

The programmable divider is not capable of dividing the 37-MHz VCO frequency, so an offset crystal-controlled oscillator is used to mix with the VCO frequency to yield a difference frequency which the divider can handle. This difference frequency maintains proper frequency relationships with the VCO frequency. If the VCO frequency increases, so will the offset frequency. Likewise, if the VCO frequency decreases, so will the offset frequency. Futhermore, the frequency changes of the VCO will be equalled by the changes in the offset frequency. For example, if the VCO frequency increases by 10 kHz, then the offset frequency will also increase by 10 kHz.

It would be easier to go through the analysis after first assuming that the radio has already been switched on and the PLL system stabilized. It only takes a fraction of a second for the PLL to stabilize after power is applied but what happens in that fraction of a second is important. The sequency of events could go like this. When power is first applied, the VCO will not operate exactly on the proper frequency. Instead, it will be operating at a random frequency. Say the VCO frequency initially starts at a lower frequency than needed. This signal is fed to the *offset mixer* where it is mixed with the signal from the *offset oscillator* to yield a difference frequency. This difference frequency is then sent to the programmable divider. So the output of the *programmable divider* would be the difference frequency divided by the programmed divisor, N. This output signal is sent to the *phase detector* where it is compared to the standard reference. The large difference in frequency causes the *phase detector* to issue a proportionately large error signal. This pulsed error signal is fed to the charge pump which in turn feeds a signal to the filter. The output of the filter is a smooth DC voltage whose amplitude is proportional to the phase or frequency difference of the phase detector inputs. This correction voltage is now applied to the control line of the VCO. Remember, changing the reverse bias to the varactor diode will change the VCO frequency. This is precisely what the correction voltage does. This causes the VCO to move its frequency in a direction which will make the output of the programmable divider approach the reference frequency. So the VCO frequency gradually increases toward the desired frequency. As the VCO approaches this frequency, the output of the programmable divider will approach the 10-kHz reference frequency. Thus, the correction voltage will become smaller and smaller. And finally, after a fraction of a second, the PLL will stabilize. At the point of stabilization, the VCO frequency will be the desired one. Because the output of the programmable divider is now equal to the reference frequency, the phase detector will issue no error signal.

There will be a voltage applied to the control line of the VCO even under locked conditions, but this voltage will not contain an error signal component. This is important to keep in mind. A study of the graph in Fig. 1-7 should clarify this point for you. The vertical axis represents the varactor reverse bias voltage, and the horizontal axis represents the VCO frequency. You can see from the graph that as the varactor reverse bias increases, the VCO frequency also increases. Conversely, when the varactor reverse bias decreases, the VCO frequency decreases. A specific bias voltage relates to a specific VCO frequency. For example, the dotted lines on the graph shows that approximately 2.2 volts bias causes the VCO frequency to be 37.660 MHz.

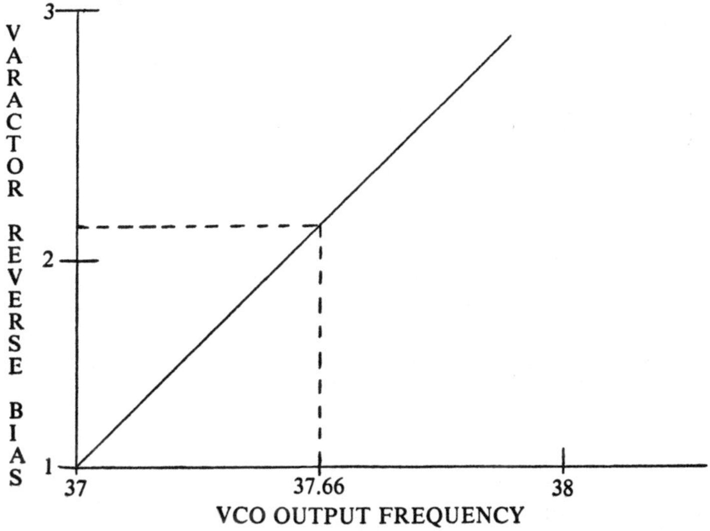

Fig. 1-7. This graph shows the relationship between the varactor bias and the vco output frequency, the dotted line shows that approximately 2.2 volts bias relates to a VCO frequency of 37.660 MHz.

SOME PRACTICAL CONSIDERATIONS

PLL synthesizers are normally very reliable and the frequencies they generate are usually very precise. Even the best of systems can falter at times, though. Lets look at a few examples of PLL malfunctions which can cause the frequencies it generates to be out of tolerance. During this discussion, refer again to Fig. 1-6. Because the reference oscillator is used as the frequency standard let's first take a look at what happens if this 10.240 MHz oscillator is operating off frequency. You might be a little surprised at the effect!

Table 1-3. Program Codes Applied to Program LInes of BCD Device of Fig. 1-2.

	Tens				Units			
Channel	P3'	P2'	P1'	P0'	P3	P2	P1	P0
1	0	0	0	0	0	0	0	1
2	0	0	0	0	0	0	1	0
3	0	0	0	0	0	0	1	1
4	0	0	0	0	0	1	0	0
5	0	0	0	0	0	1	0	1
6	0	0	0	0	0	1	1	0
7	0	0	0	0	0	1	1	1
8	0	0	0	0	1	0	0	0
9	0	0	0	0	1	0	0	1
10	0	0	0	1	0	0	0	0
11	0	0	0	1	0	0	0	1
12	0	0	0	1	0	0	1	0
13	0	0	0	1	0	0	1	1
14	0	0	0	1	0	1	0	0
15	0	0	0	1	0	1	0	1
16	0	0	0	1	0	1	1	0
17	0	0	0	1	0	1	1	1
18	0	0	0	1	1	0	0	0
19	0	0	0	1	1	0	0	1
20	0	0	1	0	0	0	0	0
21	0	0	1	0	0	0	0	1
22	0	0	1	0	0	0	1	0
23	0	0	1	0	0	0	1	1
24	0	0	1	0	0	1	0	0
25	0	0	1	0	0	1	0	1
26	0	0	1	0	0	1	1	0
27	0	0	1	0	0	1	1	1
28	0	0	1	0	1	0	0	0
29	0	0	1	0	1	0	0	1
30	0	0	1	1	0	0	0	0
31	0	0	1	1	0	0	0	1
32	0	0	1	1	0	0	1	0
33	0	0	1	1	0	0	1	1
34	0	0	1	1	0	1	0	0
35	0	0	1	1	0	1	0	1
36	0	0	1	1	0	1	1	0
37	0	0	1	1	0	1	1	1
38	0	0	1	1	1	0	0	0
39	0	0	1	1	1	0	0	1
40	0	1	0	0	0	0	0	0

Assume that the reference oscillator is operating 1000 H off frequency to the high side. This frequency would therefore be 10.241 MHz. The reference frequency would then be 10.241 MHz+1024, or 10,000.976. So a 1000-Hz increase in the reference oscillator frequency results in less than a 1-Hz increase in the basic reference frequency (0.976 Hz, to be exact). This new reference frequency will change the receive and transmit frequencies, but how much? To find out work your way backward from the output of the programmable divider to the VCO itself.

The output of the programmable divider will have to equal the new reference frequency (10,000.976 Hz) for the loop to achieve lockup. Let's say the divider is programmed to divide its input by 128. So if the output of the programmable divider is to be 10,000.976 Hz, the input to the divider will have to be 128×10,000. 976, or 1,280,124.9 Hz. This is the new offset frequency. You probably remember from the previous discussion that the offset frequency changes are directly related to VCO frequency changes. Because the new offset frequency increased by 124.9 Hz, the VCO frequency also increased by 124.9 Hz. To find the VCO frequency, add the offset frequency to the frequency of the offset oscillator. This would be 1,280,124.9 Hz÷36,380,000, or 37,660,124.9 Hz. The normal output frequency of the VCO is 37,660,000 Hz, so the VCO is operating only 124.9 Hz too high. This difference will also show up in the transmit and receive frequencies.

The interesting point here is that even though the reference oscillator frequency changed by 1000 Hz, the VCO frequency only changed by 124.9 Hz. The change in VCO frequency is equal to the change in reference frequency multiplied by the N number programmed into the divider. The formula could be written out as: Δ Fvco=Δ Fref×N. The main point to remember is that a change is in the reference oscillator frequency will also change the receive and transmit frequencies, but to a lesser degree.

If the frequency of the offset oscillator changes, then the VCO frequency will also change and by the same amount. Hence, the receive and transmit frequencies will change accordingly. If the transmit oscillator changes in frequency only, the transmit frequency will be affected.

By now you should have a basic understanding of PLL operation. The basic stages were covered individually and then analyzed as a system. The following chapters cover specific PLLs in detail and should serve to strengthen your understanding of PLLs in general.

Chapter 2
The μPD2816C Digital Synthesizer

The μPD2816C is a CMOS LSI (large-scale integrated circuit) designed especially for CB PLL synthesizers. The IC is packaged in a 22-pin dual in-line package. First we will take a close look at the muPD2816C IC itself. A block diagram of a typical application will then be analyzed in detail. Then the Cobra 1000 GTL PLL synthesizer, which uses the muPD2816C, will be explained in detail. Adjustment and test procedures will be included also.

SPECIAL FEATURES

- ☐ Only one crystal required for CB 40-channel AM operation.
- ☐ On-chip filter amplifier.
- ☐ Buffered 10.240-MHz output.
- ☐ Protection circuit for misprogramming and unlocked conditions.
- ☐ BCD 6-bit input channel select code.
- ☐ High-speed, low-power consumption due to CMOS.
- ☐ Selectable reference frequency (5 kHz or 10kHz).
- ☐ On-chip pullup resistors.
- ☐ On-chip inverter which can be used for oscillator with external crystal.

INTERNAL STRUCTURE

Figure 2-1A is a block diagram of the internal structure of the muPD2816C. The functions of the terminal pins are listed with the illustration. Pins 1 through 6 are the programming inputs, P1 through P6. The programming is applied to these pins in BCD form. Pullup resistors are shown connected to the various program pins. These resistors are part of the IC itself. If one of the program pins is ungrounded, the pullup resistor places a HIGH level (binary 1) on the pin. This simplifies the design of the channel selector switch to some extent. The program lines connect internally to the code converter. The code converter converts the BCD code to another code which is applied through the switch to the programmable divider. This then sets the divider to divide by the appropriate divisor N.

Pin 7 is the input to a divider which divides by 2. Pin 8 is the output of this divider. Pin 9 connects to the internal program switch. If pin 9 is open or at a HIGH level, the programmable divider will divide by N. If pin 9 is at ground potential, the programmable divider will divide by N + 91. The reason for this setup is that with this PLL system, the VCO frequency is not the same for the transmit and receive modes of operation. Pin 10 is connected to the output of a divider which divides the 10.240-MHz reference oscillator frequency by 2. Thus, the output of pin 10 is a 5.12-MHz signal. Pin 11 is connected to the supply voltage to power the IC. Pin 12 is the output of the 10.240-MHz oscillator. Notice that the output of the reference oscillator also feeds the input of a divider which divides it by 2. The output of the divide by 2 divider is then 5.12 MHz. This is then divided by the divide by 512 divider to yield 10 kHz. This 10-kHz signal can be used for the reference frequency for the phase detector or if a 5-kHz reference frequency is desired, the reference input of the phase detector is taken from the output of another divide by 2 divider, which divides the 10-kHz signal by 2 to yield 5 kHz. The reference input to the phase detector is determined by an electronic switch. This switch is drawn as a mechanical switch in the illustration for simplification. The voltage level applied to pin 20 determines whether the switch will connect the phase detector reference input to the 5-kHz or the 10-kHz terminal. If pin 20 is connected to a HIGH level or simply left open, the reference frequency will be 10 kHz. If pin 20 is grounded, the reference frequency will be 5 kHz.

Pin 14 is a buffered 10.240-MHz output. Pin 15 serves to stop the transmitter if an unlocked condition exists. Pin 16 is used to connect an external RC network and supply voltage. The output of the lock detector is a series of pulses which are fed to the external integrator circuit through the MOS transistor. The integrated signal is then used to set the level at pin 15. Pin 17 is the error signal output from the charge pump. This error signal can be applied through external components to the input of the active filter amplifier at pin 18. The output of the active filter amplifier is brought out to pin 19. This DC voltage can then be used to control the VCO. Pin 21 is common or ground. Pin 22 is the input to the programmable divider. The pins are numbered as shown in Fig. 2-1B.

Table 2-1 shows the 40-channel codes that are applied to the programming inputs of the μPD2816C. Also shown at the bottom of the table are six input codes which are *not allowed.* Program lines P6 and P5 are used to represent the tens digit, and program lines P4, P3, P2, and P1 are used to represent the units digit. The weight of P6 is 2 and the weight of P5 is 1. The weights of P4 through P1 are P4=8, P3=4, P2=2, and P1=1. To find the decimal equivalent of the code add the weights of the pins where binary 1 appears. For example for channel 37 to get the tens position digit, P6 and P5 are at binary 1 so add their weights together:

P6 (weight) = 2
P5 (weight) = 1

Tens digit = 3

P3, P2, and P1 are at binary 1 so add the weights of these positions to get the units digit.

P1 (weight) = 1
P2 (weight) = 2
P3 (weight) = 4

Units digit = 7

So we have the tens digit of 3 and the units digit of 7. One exception to this is channel 40. On channel 40, all the program lines are LOW (binary 0). The code converter recognizes this as channel 40. This simplifies the programming.

Notice at the bottom of the chart that there are six program codes for P4 through P1 which are not allowed. If these codes appear on the program lines P4 through P1, the level at the lock detector output (IS) will go low. This is used to kill the transmitter to prevent out-of-band signal radiation.

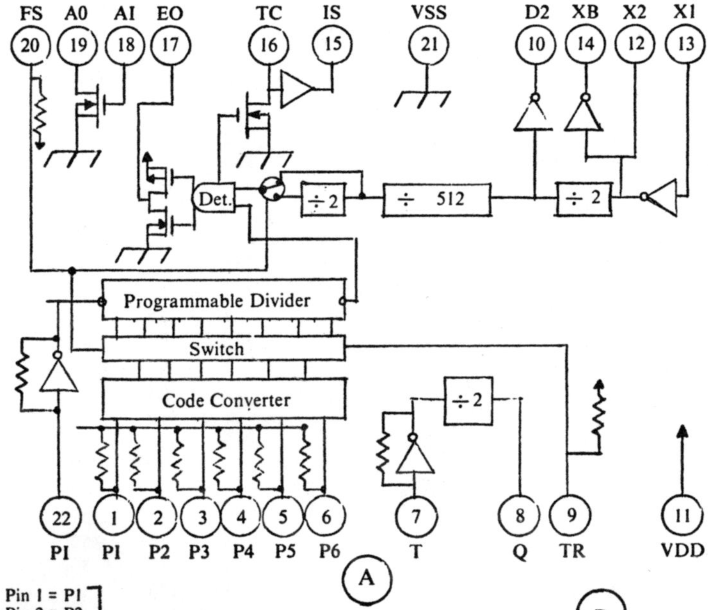

Pin 1 = P1
Pin 2 = P2
Pin 3 = P3
Pin 4 = P4
Pin 5 = P5
Pin 6 = P6 } Program Inputs
Pin 7 = ½ T-F/F Input
Pin 8 = ½ T-F/F Output
Pin 9 = TX/RX Switch Input (high level or open RX mode; low level or ground TX mode))
Pin 10 = 5.12 MHz output
Pin 11 = Power supply voltage (VDD)
Pin 12 = Crystal oscillator output
Pin 13 = Crystal oscillator input } 10.240 Mhz
Pin 14 = Buffered 10.24 MHz output
Pin 15 = Lock detector output
Pin 16 = Lock detector integrator
Pin 17 = Error signal output
Pin 18 = Filter amplifier input
Pin 19 = Filter amplifier output
Pin 20 = Reference frequency switch input (high level = 10 kHz. low level or ground = 5 kHz)
Pin 21 = Ground (VSS)
Pin 22 = Programmable divider input

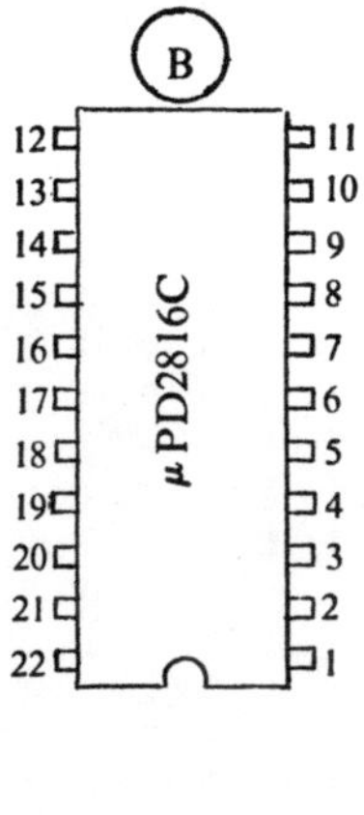

Fig. 2-1. Block diagram of the internal structure of the μ PD2816C is shown at A. The drawing at B is a top view of the μ PD2816C showing pin numbering. The functions of the various pins are listed.

Table 2-1. Program Codes, VCO Frequencies, and Divisors.

Chan.	Program Input P6	P5	P4 (Units)	P3	P2	P1	TR = 1(Rx) FS = 0 (5 kHz) N	VCO Fr.	TR = 0(Tx) FS = 0 (5 kHz) N	VCO Fr.	TR = 1 FS = 1 (10 kHz) N	VCO Fr.	IS
1	0	0	0	0	0	1	182	16.27 MHz	273	16.725 MHz	91	16.27 MHz	1
2	0	0	0	0	1	0	184	.28	275	.735	92	.28	1
3	0	0	0	0	1	1	186	.29	277	.745	93	.29	1
4	0	0	0	1	0	0	190	.31	281	.765	95	.31	1
5	0	0	0	1	0	1	192	.32	283	.775	96	.32	1
6	0	0	0	1	1	0	194	.33	285	.785	97	.33	1
7	0	0	0	1	1	1	196	.34	287	.795	98	.34	1
8	0	0	1	0	0	0	200	.36	291	.815	100	.36	1
9	0	0	1	0	0	1	202	.37	293	.825	101	.37	1
10	0	1	0	0	0	0	204	.38	295	.835	102	.38	1
11	0	1	0	0	0	1	206	.39	297	.845	103	.39	1
12	0	1	0	0	1	0	210	.41	301	.865	105	.41	1
13	0	1	0	0	1	1	212	.42	303	.875	106	.42	1
14	0	1	0	1	0	0	214	.43	305	.885	107	.43	1
15	0	1	0	1	0	1	216	.44	307	.895	108	.44	1
16	0	1	0	1	1	0	220	.46	311	.915	110	.46	1
17	0	1	0	1	1	1	222	.47	313	.925	111	.47	1
18	0	1	1	0	0	0	224	.48	315	.935	112	.48	1
19	0	1	1	0	0	1	226	.49	317	.945	113	.49	1
20	1	0	0	0	0	0	230	.51	321	.965	115	.51	1
21	1	0	0	0	0	1	232	.52	323	.975	116	.52	1
22	1	0	0	0	1	0	234	.53	325	.985	117	.53	1
23	1	0	0	0	1	1	240	.56	331	17.015	120	.56	1
24	1	0	0	1	0	0	236	.54	327	16.995	118	.54	1
25	1	0	0	1	0	1	238	.55	329	17.005	119	.55	1
26	1	0	0	1	1	0	242	.57	333	.025	121	.57	1
27	1	0	0	1	1	1	244	.58	335	.035	122	.58	1
28	1	0	1	0	0	0	246	.59	337	.045	123	.59	1
29	1	0	1	0	0	1	248	.60	339	.055	124	.60	1
30	1	1	0	0	0	0	250	.61	341	.065	125	.61	1
31	1	1	0	0	0	1	252	.62	343	.075	126	.62	1
32	1	1	0	0	1	0	254	.63	345	.085	127	.63	1
33	1	1	0	0	1	1	256	.64	347	.095	128	.64	1
34	1	1	0	1	0	0	258	.65	349	.105	129	.65	1
35	1	1	0	1	0	1	260	.66	351	.115	130	.66	1
36	1	1	0	1	1	0	262	.67	353	.125	131	.67	1
37	1	1	0	1	1	1	264	.68	355	.135	132	.68	1
38	1	1	1	0	0	0	266	.69	357	.145	133	.69	1
39	1	1	1	0	0	1	268	.70	359	.155	134	.70	1
40	0	0	0	0	0	0	270	.71	361	.165	135	.71	1
			1	0	1	0							0
			1	0	1	0							0
			1	1	0	0							0
			1	1	0	1							0
			1	1	1	0							0
			1	1	1	1							0

0 = low level (binary 0)
1 = high level (binary 1)

Notice that the first column (where TR=1 and FS=0) lists the N numbers and the VCO frequencies. When TR=1, this is the receive mode. The transmit mode is when TR=0. Where FS=0, the reference frequency is 5 kHz. If FS = 1, the reference frequency is 10 kHz. The first column lists the N numbers (divisors) and the VCO frequencies for the receive mode where the reference frequency is 5 kHz. The third column lists the N divisors and the VCO frequencies where the reference is 10 kHz (receive mode).

Notice that where 5 kHz is used as the reference frequency the N numbers increase by 2 for each incremental channel change. This is because the CB channel spacing is 10 kHz. Because the reference frequency is only half that frequency, the N will have to be changed by 2 when switching between channels separated by 10 kHz. The spacing between channels 3 and 4 is 20 kHz, so the N number would have to change by 4 when switching between 3 and 4. There is also a 20-kHz separation between CB channels 7 and 8, 11 and 12, 15 and 16, and 19 and 20. You will notice on the chart that the N divisors change by 4 when switching between these channels. Also, between channels 22 and 23 there is a 30-kHz separation, so switching between these channels requires the N number to change by 6. The new channels, 24 and 25, were placed between 22 and 23. You can see in column 1 of Table 2-1 that the N divisor for channel 22 is 234 and for channel 23 is 240. On channel 24, the N number goes back to 236 and for channel 25 is 238. There is another 20-kHz jump between channels 25 and 26, but channel 23 fits between channels 25 and 26.

The third column in Table 2-1 is based on a reference frequency of 10 kHz. The column shows the various N values and VCO frequencies for the receive mode. Notice that the N numbers are consecutive, except where odd channel spacing exists. This is because the reference frequency is the same as the channel spacing -10 kHz.

A TYPICAL APPLICATION

Figure 2-2 shows a block diagram of a theoretical application of the μ PD2816C. Let's analyze the operation of the system using channel 23 as the example. The program lines (P1 through P6) are connected to switches that can either ground the program line or leave it open. The line is at binary 0 when grounded. If the line is left open, the internal pullup resistor will place the line at a HIGH level or binary 1. The switches are shown as being set up for channel 23. And for channel 23, the code converter sets the divider

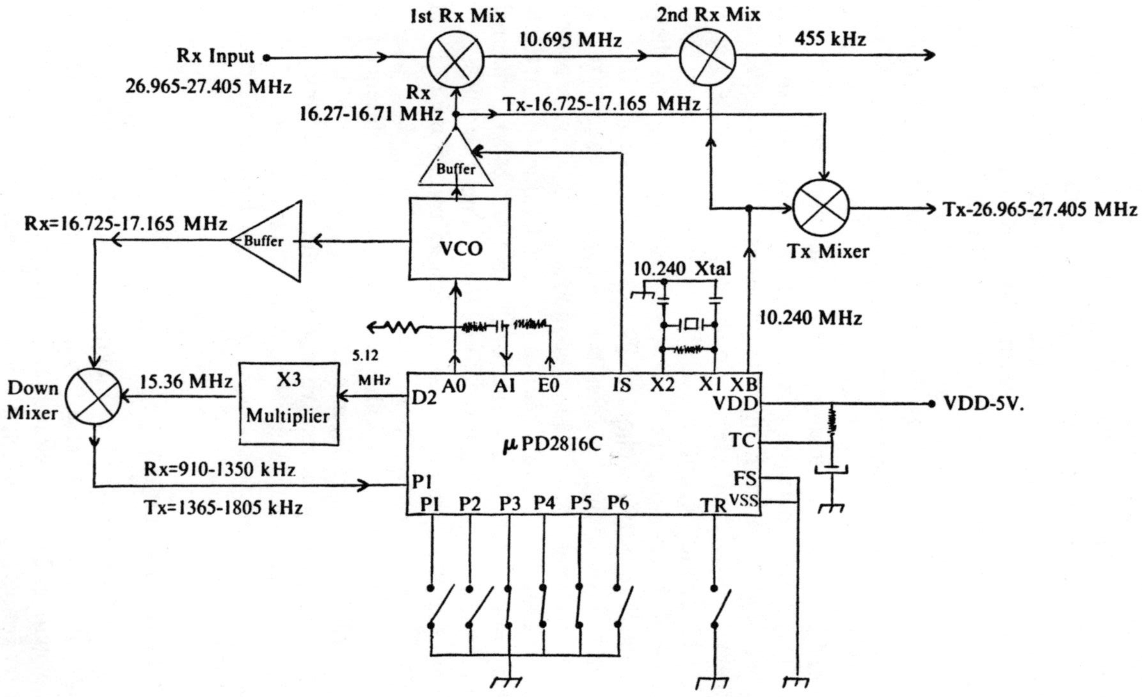

Fig. 2-2. This is a block diagram of a typical application of the μ PD2816C.

to divide by 240 in the receive mode. For the transmit mode, the divider will be set up to divide by 331. Notice that FS is grounded so this sets the reference frequency at 5 kHz. In the receive mode, the VCO frequency is mixed with the incoming receive signal to produce a difference frequency of 10.695 MHz. This is the first intermediate frequency. This 10.695-MHz i-f is then sent to the *second mixer* where it is mixed with the signal from the buffered 10.240-MHz output (XB) of the IC. The difference of these two signals (10.695-10.240 = 455 kHz) is the low intermediate frequency. In the transmit mode, the VCO frequency is mixed with a 10.240 MHz signal (from the XB terminal of the IC) to produce a sum frequency that is the transmit signal.

The VCO frequency is mixed with a 15.360-MHz signal to produce a difference output which the programmable divider can handle. Notice the switch connected to the TR terminal of the IC. In the receive mode, the switch is open and the divider is set to divide by N. When the switch is closed, TR is grounded. This changes the divisor to N + 91.

The frequency of 27.255 MHz is CB channel 23. To receive this signal, the VCO frequency will have to be 27.255-10.695 (first i-f), or 16.560 MHz (see Table 2-1). The VCO frequency is mixed with a 15.360-MHz signal in the down mixer to produce the frequency 16.560-15.360, or 1.200 MHz. This 1.200-MHz signal is sent to the programmable divider where it is divided by 240 to yield 5 kHz. Because the output of the divider is equal to the reference frequency, no correction voltage will be issued so the VCO frequency remains constant. Notice that the error signal output, EO, is fed to the input of the filter amplifier, AI. The control voltage for the VCO is taken from the output of the filter amplifier, AO.

In the transmit mode, the TR terminal of the IC is grounded when the transmitter is keyed. This can be done through the microphone switch, through relay contacts, or by an electronic switch. Regardless of how this is achieved, when TR is grounded, the divisor of the programmable divider is increased by 91. In the channel 23 receive mode, the divisor is 240. In the transmit mode, therefore, the divisor becomes 240+91, or 331. At the instant the transmitter is keyed, the VCO frequency will still be 16.560 MHz. Therefore, the input to the divider will be 1.200 MHz ÷ 331, or 3.625 kHz. Thus, a large error now exists between the divider output frequency and the 5-kHz reference frequency. The phase detector detects this error and produces a corresponding error signal. This error signal is amplified and filtered by the filter amplifier and then applied to the control line of the VCO. This error signal causes the VCO frequency to gradually increase. As the VCO frequency increases, the output of the programmable divider will move closer and closer to 5 kHz. When the output of the programmable divider reaches 5 kHz, the phase detector produces no more error signal. Hence, the VCO frequency stabilizes. The new VCO frequency is now 17.015 - 15.360, or 1.655 MHz. This is in turn divided by 331 to yield 5 kHz.

To obtain the transmit frequency, the VCO frequency is mixed with the 10.240-MHz signal: 17.015 MHz + 10.240 MHz = 27.255 MHz. When the transmitter is first keyed, a finite amount of time is required for the VCO frequency to reach and stabilize at the new frequency. During this time interval, the PLL is unlocked. The lock detector detects this unlocked condition and issues a signal that kills the transmitter output. In Fig. 2-2, this signal is fed from IS to the buffer to kill the buffer stage. This in turn kills the transmitter output. This prevents improper signal radiation from the transmitter. When the VCO reaches the correct frequency, the PLL will lock. The lock detector output will then return to normal and thus enable the transmitter. Actually, it only requires a small fraction of a second for the PLL to lock.

When the transmitter is unkeyed, the programmable divider will revert to the divisor 240. Momentarily, the VCO will still be operating at 17.015 MHz. The input to the divider will still be 1.655 MHz. The output of the divider will be 1.655 MHz ÷ 240, or 6.896 kHz. This is quite higher than the 5-kHz reference, and the phase detector detects this and produces an error signal that is filtered and then applied to the control line of the VCO. This error signal causes the VCO frequency to decrease. As it decreases, the output of the programmable divider approaches the 5-kHz reference frequency. When the output frequency of the divider reaches 5 kHz, the phase detector produces no more error signal, so the VCO stabilizes. The VCO is now back at 16.560 MHz.

COBRA 1000 GTL CB SYNTHESIZER

The Cobra 1000 GTL is pictured in Fig. 2-3. The digital synthesizer circuit of the Cobra 1000 GTL is built around the μ PD2816C IC. During this discussion, refer to the schematic in Fig. 2-4. Also, Table 2-1 will be referred to in this discussion.

Remember that the μPD2816C can have a reference frequency of 5 kHz or 10 kHz, depending on the voltage level at pin 20. If pin 20 is grounded, the reference frequency is 5 kHz. If pin 20

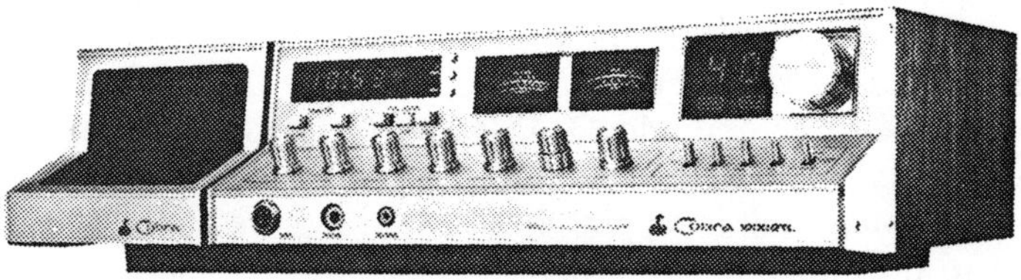

Fig. 2-3. The Cobra 1000 GTL Transceiver.

is high, the reference frequency is 10 kHz. On the schematic, pin 20 is connected to ground, so this sets the reference frequency at 5 kHz. The reference oscillator operates at 10.240 MHz. This oscillator consists of transistor TR14 and the associated circuitry. The frequency of the oscillator is determined primarily by crystal X1 which resonates at 10.240 MHz. Inductor L21 is used to set the oscillator precisely on frequency.

Also, notice varactor diode D18. This is shown connected from one side of crystal X1 to ground. The varactor serves as a fine tuning control. In the receive mode, diode D19 is forward biased. This allows the voltage at the arm of the delta-tune control (VR306) to appear across the varactor diode, D18. This voltage reverse biases the varactor. As the delta-tune control is varied, the reverse bias across D18 also varies. Hence, the capacitance exhibited by D18 varies. The changing capacitance of D18 will change the oscillator frequency. This can improve the reception of off-frequency signals. In the transmit mode, it is undesirable for the delta-tune control to affect the oscillator frequency. Otherwise, the transmitter would be operating off frequency. This is illegal, of course! So in the transmit mode, 7.9 volts is applied to the anode of diode D20. This sets up a current flow from ground through R80, R84, and D20 back to the 7.9-volt supply. Resistors R80 and R84 form a voltage divider. The voltage at the junction of R80 and R84 will be approximately 3.5 volts. This will set a fixed reverse bias across the varactor. Also, the supply voltage in the transmit mode is removed from the delta-tune control, so D19 will be reverse biased. This isolates the delta-tune control from the circuit.

The 10.240-MHz signal that is developed by the reference oscillator is fed to pin 13 of IC4. This signal is divided by 2048 to yield a 5-kHz reference signal at the reference input of the phase detector.

The voltage-controlled oscillator consists of part of IC2, along with the external components. The components that determine the resonant frequency of the VCO are L20, C94, and, of course, the varactor diode, D14. The VCO designed to operate in the 16 to 17 MHz range. The exact frequency is determined by the control voltage fed from pin 19 of IC4 through resistors R70 and R71 to the anode of the varactor, D14.

From a previous discussion on the μPD2816C, you learned that a 5.12-MHz signal is available at pin 10 of IC4. Figure 2-4 shows that L19 is connected to pin 10 through capacitor C80. The primary of L19 is tuned to 15.360 MHz (the third harmonic of 5.12 MHz). The secondary of L19 feeds this 15.360-MHz signal to one input of the VCO/mixer, pin 4 of IC2. Inside IC2, the 15.360-MHz signal is mixed with the VCO frequency to yield FVCO - 15.360 MHz, or FPDI, where FVCO is the VCO frequency and FPDI is the programmable divider input frequency.

The operation of the entire PLL circuit using channel 1 as the example, is next. In the receive mode on channel 1, the programmable divider is set to divide its input frequency by 182. To yield a 5-kHz signal at the output of the programmable divider, the input to the divider will have to be 5000 × 182, or 910,000 Hz or 0.910 MHz. By the formula, then, the VCO frequency would have to be FVCO = 15.360 + FPDI = 15.360 + 0.910, or 16.270 MHz. The difference frequency of 0.910 MHz is taken from pin 6 of IC2 and fed to the input of the programmable divider at pin 22 of IC4. The frequency at the input of the programmable divider varies directly as the VCO frequency varies. On channel 1, if the input to the programmable divider is not exactly 0.910 MHz, an error signal will appear at the output of the phase detector. This error signal will be amplified, filtered, and then applied to pin 19 of IC4. This error signal will change the control voltage applied across varactor diode D14. This will in turn move the VCO frequency until the input to the divider is exactly 0.910 MHz. When the input to the divider is exactly 0.910 MHz, the phase detector will produce no more error signal. Thus, the control voltage stabilizes at a level that keeps the VCO at 16.270 MHz and consequently the input to the programmable divider at 0.910 MHz.

The VCO frequency is taken from pin 3 of IC2. From pin 3,

the signal is fed through L18 to the receiver first mixer where it is heterodyned with the incoming channel 1 signal to yield 26.965 MHz - 16.270 MHz, or 10.695 MHz. This 10.695-MHz signal is the first i-f. This first i-f signal is then mixed with a 10.240-MHz signal which is taken from the secondary of L23. The difference frequency is 10.695 MHz - 10.240 MHz, or 455 kHz. This is the second intermediate frequency.

In the transmit mode, the VCO is made to operate at a frequency which is 455 kHz higher than the receive mode frequency. For channel 1, the VCO frequency would therefore be 16.725 MHz. Let's see how this change in VCO frequency is accomplished. When the transmitter is keyed, a set of contacts on the push-to-talk switch is used to ground pin 9 of IC4 through a 2.2-ohm resistor (R103) and diode D17. When pin 9 of IC4 goes low, the divisor of the programmable divider increases by 91. The divisor in the receive mode was 182, so the divisor for the transmit mode will be 182 + 91, or 273. So the input to the programmable divider must be 5000 × 273, or 1.365 MHz. Notice that this is a 455-kHz increase over the receive mode divider input frequency.

When the transmitter is first keyed, a large difference will exist in the output frequency of the programmable divider and the 5-kHz reference frequency. The phase detector issues a large error signal. As a result, a correction voltage is applied to the VCO control line. This causes the VCO frequency to move upward until the frequency at the input of the programmable divider reaches 1.365 MHz. When the input to the divider is 1.365 MHz, the output of the programmable divider will be 5 kHz. Because this is the same as the reference frequency, the phase detector will not produce an error signal now. This allows the voltage on the control line of the VCO to stabilize the vco at 1.365 + 15.360, or 16.725 MHz.

A sample of the VCO frequency is taken from the secondary of L20 and fed to pin 1 of IC3 (transmit mixer). Also, a 10.240 MHz signal from the reference oscillator is fed to pin 4 of IC3. The sum of these two frequencies is taken from pin 9. The frequency at pin 9 for channel 1 is 10.240 MHz + 16.725 MHz, or 26.965 MHz. This signal is fed through L25 on to the buffer stage.

When an out-of-lock condition exists, pin 15 of IC4 will go LOW. This will kill transmit buffer stage TR12. Also, a misprogramming condition will cause the lock detector output to go LOW. By measuring the voltage at pin 15, you can immediately tell whether the PLL is locked or unlocked. This is an important troubleshooting clue.

Alignment

☐ With an rf voltmeter connected to TP6, adjust L23 for maximum indication on the meter. Make this adjustment with the channel selector on channel 20.

☐ Connect a frequency counter to pin 13 of IC4 and check the accuracy of the 10.240-MHz reference frequency. Make sure that the delta-tune control is in the center or detent position. Also, check this frequency in the transmit mode. There should be very little difference in frequency in the transmit and receive modes of operation.

☐ With the set on channel 40, monitor the DC voltage at TP4. Adjust L20 for 2.1 volts. On channel 1, the voltage should be approximately 1.3 volts. Make this adjustment very carefully, because this is a fairly critical adjustment.

☐ With an rf voltmeter connected to pin 4 of IC2, adjust L19 for maximum indication on the meter. Then, using a frequency counter, check for 15.360 MHz at pin 4 of IC2.

☐ With a frequency counter connected to TP5 (pin 22 of IC4), check on all channels for the frequencies shown in Table 2-2. Check this in the receive and transmit modes of operation.

☐ With an rf voltmeter connected to TP2, adjust L18 for maximum indication on the meter.

☐ Using a frequency counter, check TP2 for the proper VCO frequencies on all channels. Check this in the receive and the transmit modes. Refer to Table 2-1 for the proper frequencies for all 40 channels.

☐ With an rf voltmeter connected to TP1, adjust 124 and 125 for maximum indication on the meter.

☐ Using a counter, check for the proper transmit frequencies at the transmitter output.

Troubleshooting Hints

There are two test points that tell quite a lot about the condition of the PLL. These are the lock detector/misprogramming detector output and the VCO control line voltage at TP4.

If the voltage at pin 15 of IC4 is normal, this indicates that the PLL is locked and operating. But the VCO frequency could still be wrong even with a locked condition. For example, if the 15.360-MHz signal were off frequency, the VCO would be forced to operate off frequency to maintain the proper input frequency at the programmable divider. If the voltage at pin 15 of IC4 is low, this is an indication that either the loop is unlocked or that improper programming is applied to the program inputs.

If the control line voltage at TP4 is abnormally high or low, this indicates that the VCO frequency is wrong or missing altogether. This could also indicate that the reference (10.240-MHz) signal is wrong or missing. Also if the 15.360-MHz signal is wrong or missing, the voltage at TP4 will be abnormal. Infact, almost any trouble within the loop will affect the voltage at TP4, so you can see how important it is to check here first when troubleshooting this PLL circuit.

No Transmit or Receive On Any Channel. Check for the presence of rf voltage at TP2 (output of VCO). If there is no rf voltage at TP2, troubleshoot the VCO. If there is rf voltage at TP2 check the frequency with a frequency counter just to verify that it is in the 16 to 17 MHz range. If so, check for the 15.360-MHz signal at TP8. If this frequency is wrong or missing, check the reference oscillator signal at TP7. If the 15.360-MHz signal is present at TP8, check for rf voltage at TP5. The frequency at TP5 should be equal to the VCO frequency minus 15.360 MHz. Check to see that it is. If not, check the VCO/mixer stage.

Providing everything else checks OK, a VCO adjustment might be all that is needed. If L20 is badly out of adjustment, the voltage at TP4 will be either abnormally high (near 4.8 volts) or abnormally low (near 0.5 volts). With the set on channel 1, monitor the DC voltage at TP4 while carefully adjusting L20. If, after a few turns in one direction, the voltage at TP4 does not change, turn the slug in the opposite direction. After a few turns or less, the voltage at TP4 should start to change. When it does, adjust it to 1.3 volts on channel 1. If the voltage at TP4 does not change, then there might be a defect in the μPD2816C itself. In this case you would need to return L20 to its original setting.

No Operation On High Channels With Low Channels OK. This is almost always caused by misadjustment of the VCO. Refer to the third step under ALIGNMENT.

No Operation On Low Channels With High Channels OK. This is almost always caused by misadjustment of the VCO. Refer to the third step under ALIGNMENT.

Weak Receive, Off-Frequency Transmit On All Channels. With this condition, the VCO frequency is not correct, but the fault does not lie with the VCO itself. The 10.240-MHz oscillator is usually to blame for this problem. Suppose that the actual fre-

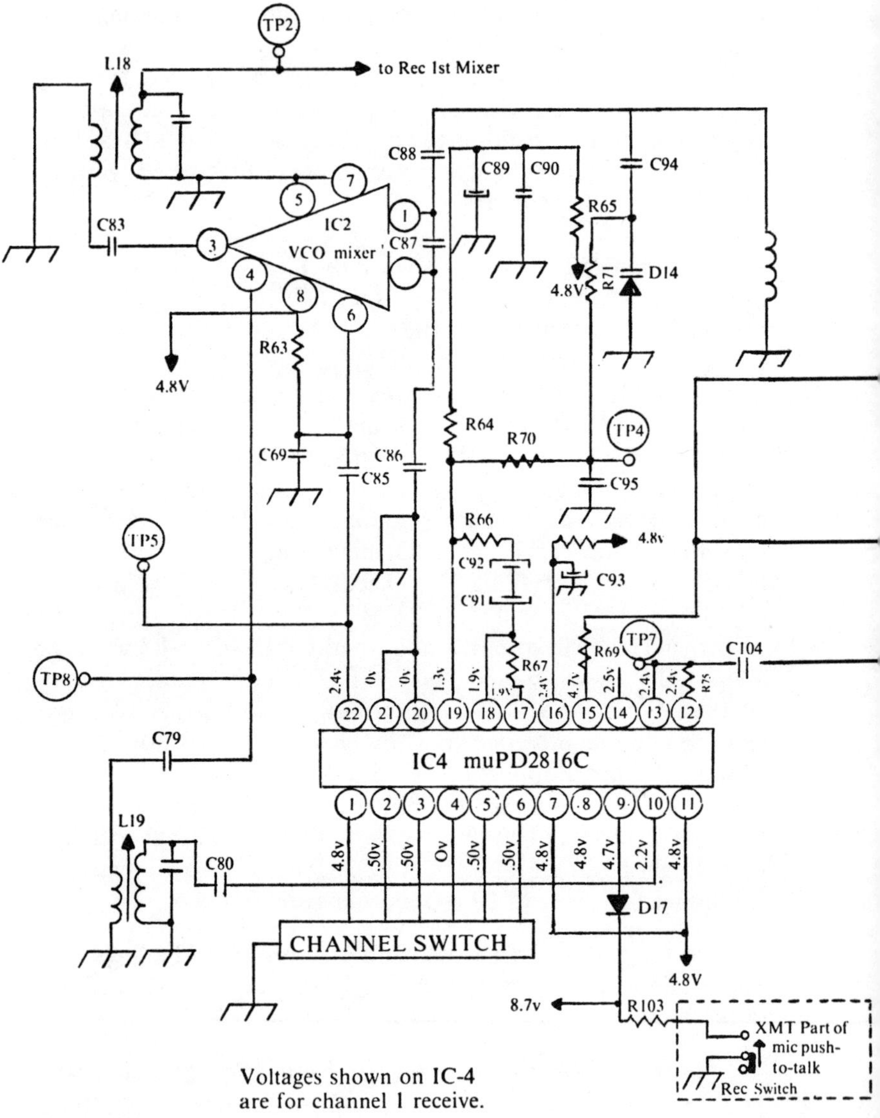

Fig. 2-4. A schematic of the PLL circuit used in the Cobra 1000 GTL.

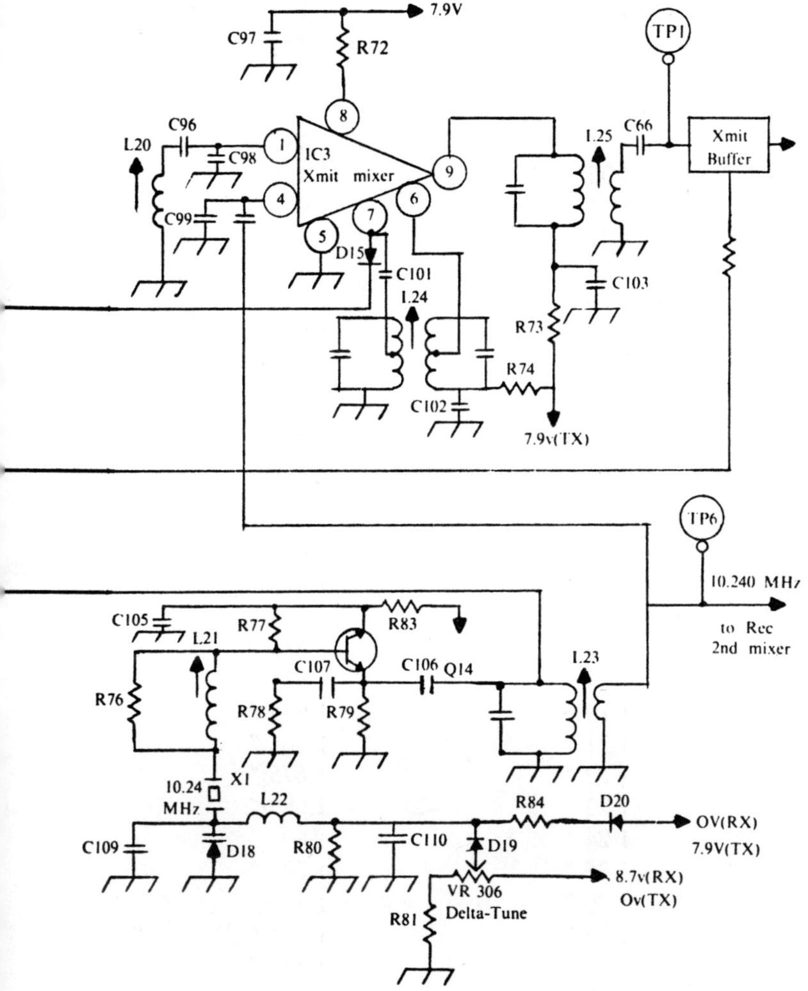
7.9V
C97
R72
TP1
C96
L20
C98
IC3
Xmit mixer
L25
C66
Xmit
Buffer
C99
D15
C101
C103
L24
R73
R74
C102
7.9v(TX)
TP6
10.240 MHz
to Rec
2nd mixer
C105
R77
R83
L21
L23
C107
C106
Q14
R76
R78
R79
X1
10.24
MHz
L22
R84
D20
OV(RX)
7.9V(TX)
C109
D18
R80
C110
D19
8.7v(RX)
Ov(TX)
VR 306
Delta-Tune
R81

Table 2-2. Correct Frequencies for Test Point 5 in the Cobra 1000 GTL PLL Synthesizer.

Channel	Frequency At TP5 RX MODE	Frequency At TP5 TX MODE
1	.91 MHz	1.365 MHz
2	.92 "	1.375 "
3	.93 "	1.385 "
4	.95 "	1.405 "
5	.96 "	1.415 "
6	.97 "	1.425 "
7	.98 "	1.435 "
8	1.00 MHz	1.455 "
9	1.01 "	1.465 "
10	1.02 "	1.475 "
11	1.03 "	1.485 "
12	1.05 "	1.505 "
13	1.06 "	1.515 "
14	1.07 "	1.525 "
15	1.08 "	1.535 "
16	1.10 "	1.555 "
17	1.11 "	1.565 "
18	1.12 "	1.575 "
19	1.13 "	1.585 "
20	1.15 "	1.605 "
21	1.16 "	1.615 "
22	1.17 "	1.625 "
23	1.20 "	1.655 "
24	1.18 "	1.635 "
25	1.19 "	1.645 "
26	1.21 "	1.665 "
27	1.22 "	1.675 "
28	1.23 "	1.685 "
29	1.24 "	1.695 "
30	1.25 "	1.705 "
31	1.26 "	1.715 "
32	1.27 "	1.725 "
33	1.28 "	1.735 "
34	1.29 "	1.745 "
35	1.30 "	1.755 "
36	1.31 "	1.765 "
37	1.32 "	1.775 "
38	1.33 "	1.785 "
39	1.34 "	1.795 "
40	1.35 "	1.805 "

quency of the reference oscillator is 10.242 MHz. This would then make the reference frequency at the phase detector 10.242 MHz ÷ 2048, or 5,000.9 Hz. So the phase detector reference frequency changed by less than 1 Hz. Obviously, this won't have much effect on the VCO frequency, but the frequency at TP8 will no longer be 15.360 MHz. Instead, it will be $\frac{10.242 \times 3}{2}$, or 15.363 MHz. This will cause the VCO frequency to operate 3000 Hz higher in order for the frequency at TP5 to remain unchanged.

When this condition is present, your first step should be to check the reference frequency at TP6 with a frequency counter. If the frequency is wrong, adjust or repair the 10.240-MHz reference oscillator.

Some Channels Inoperative. Check the voltage levels at the program pins 1 through 6 on IC4. If these levels don't correspond with those shown in Table 2-1 for the selected channel, the problem must be in the channel selector switch or a short or open on the program lines. If none of these conditions are present, the μPD2816C itself might be at fault.

Some Channels Intermittent. Check the channel selector switch. Clean it with a good tuner cleaner if needed. If necessary, replace the channel switch.

No Transmit With Receive OK. Check the transmit mixer stage (IC3). Check to see if pin 9 of IC4 is low when the transmitter is keyed. If not, check the keying circuit and diode D17.

No Receive With Transmit OK. Check for proper VCO frequency at TP2 and at the first mixer stage. Check the voltage at pin 9 of IC4. It should be around 4.7 volts in the receive mode.

Chapter 3
The SC42502P/ 3001-201 Digital Synthesizer

The SC42502P or 3001-201 IC is a custom made IC. It is manufactured especially for E. F. Johnson by one of the major semiconductor manufacturers. This is a CMOS device and is encased in a 16-pin dual-in-line package. This device is used in many Johnson CB models. In this discussion, the Johnson Messenger 4140 will be used as the example.

First, the IC itself will be discussed. Then a block diagram of a typical application of the IC will be analyzed. Next, a thorough analysis of the Johnson Messenger 4140 PLL synthesizer will be presented. Also, alignment and troubleshooting will be covered. Then, the LED readout display circuitry will be explained.

SPECIAL FEATURES

- ☐ Low-Power CMOS
- ☐ On-chip oscillator (with external xtal)
- ☐ On-chip pullup resistors
- ☐ Wide supply voltage range
- ☐ 16 pin dual-in-line construction
- ☐ Lock detector which can be used to disable transmitter to prevent radiation of off frequency radiation.
- ☐ Double diode protection on all inputs

A schematic of this dual-diode protection circuit is shown in

Fig. 3-1. If the input voltage rises to a high enough positive value, diode D1 will become forward biased and will then clamp the input voltage to the voltage level (+V) on the cathode of D1. If the input voltage tries to go negative, it will be clamped to ground by diode D2. This protects the device from excessive voltage spikes of either polarity.

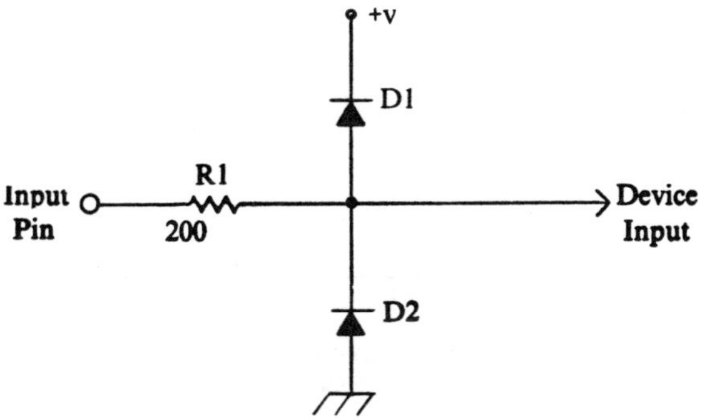

Fig. 3-1. This is the protection network used on all inputs to the SC42502P/3001-201 digital synthesizer IC. A high positive spike is clamped to (+V) through D1, while a high negative spike is clamped to ground through D2 (courtesy of E. F. Johnson Co.).

INTERNAL STRUCTURE

The block diagram in Fig. 3-2 shows a partial breakdown of the internal structure of the SC42502P/3001-201 IC. Figure 3-3 shows the pin numbering scheme of the IC. Pin 1 is the input to the programmable divider. Pins 2, 3, 4, 5, 12, 13, 14, and 15 are the *programming inputs* to the programmable divider. The programmable divider used in this IC may be either of two types. For one type, pin 6 should be at a HIGH level. For the other type, pin 6 should be at a LOW level or ground. This makes the IC compatible. Pin 7 is the lock detector output. During normal operation (under locked conditions), the voltage at pin 7 is high, around 10 volts. When an unlocked condition occurs, the voltage at pin 7 goes low. This can be used to stop the transmitter if an unlocked condition occurs. Pin 8 is the output of the *frequency/phase detector.* Pin 9 is the ground pin for the IC. Pin 10 is the output side of the CMOS oscillator/buffer, and pin 11 is the input to the CMOS oscillator/buffer. Pin 16 is connected to the supply voltage (VDD) to furnish power to the IC.

The programming codes and the divisors for channels 1-40

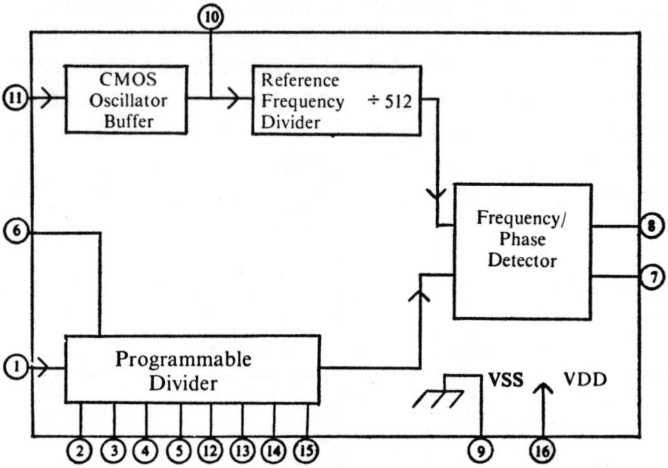

Pin 1 - Input to Programmable Divider
2, 3, 4, 5 } Program Inputs
6 - Hi on Low, depending on type of programmable divider
7 - Phase detector output
8 - Lock-Detector output
9 - VSS (ground)
10, 11 } crystal for reference osc.
12, 13, 14, 15 } Program Inputs
16 - VDD (supply voltage)

Fig. 3-2. A partial breakdown on the SC42502P/3001-201 digital synthesizer IC. The functions of the various pins are listed.

are shown in Table 3-1. The weights of the various programming pins are shown in parenthesis below the pin number. To get the divisor number, simply add the weights of the pins where binary 1 appears. For example, for channel 20 the binary 1 appears on pins 12, 14, 3, and 5. The divisor is calculated as follows:

pin 12 (weight) = 80
pin 14 (weight) = 20
pin 13 (weight) = 4
pin 5 (weight) = 1
divisor = 105

The binary level on the various program pins is determined by the channel selector switch, which either grounds or ungrounds the pin. If a pin is grounded, it is at the binary 0 level. If a pin is ungrounded, it is pulled to a HIGH level by the internal pullup resis-

Table 3-1. Divide-By Numbers and Programming Codes.

Channel NO.	Divide By NO.		Programming Code							
		Pin	12	13	14	15	2	3	4	5
		Weight	(80)	(40)	(20)	(10)	(8)	(4)	(2)	(1)
1	81		1	0	0	0	0	0	0	1
2	82		1	0	0	0	0	0	1	0
3	83		1	0	0	0	0	0	1	1
4	85		1	0	0	0	0	1	0	1
5	86		1	0	0	0	0	1	1	0
6	87		1	0	0	0	0	1	1	1
7	88		1	0	0	0	1	0	0	0
8	90		1	0	0	1	0	0	0	0
9	91		1	0	0	1	0	0	0	1
10	92		1	0	0	1	0	0	1	0
11	93		1	0	0	1	0	0	1	1
12	95		1	0	0	1	0	1	0	1
13	96		1	0	0	1	0	1	1	0
14	97		1	0	0	1	0	1	1	1
15	98		1	0	0	1	1	0	0	0
16	100		1	0	1	0	0	0	0	0
17	101		1	0	1	0	0	0	0	1
18	102		1	0	1	0	0	0	1	0
19	103		1	0	1	0	0	0	1	1
20	105		1	0	1	0	0	1	0	1
21	106		1	0	1	0	0	1	1	0
22	107		1	0	1	0	0	1	1	1
23	110		1	0	1	1	0	0	0	0
24	108		1	0	1	0	1	0	0	0
25	109		1	0	1	0	1	0	0	1
26	111		1	0	1	1	0	0	0	1
27	112		1	0	1	1	0	0	1	0
28	113		1	0	1	1	0	0	1	1
29	114		1	0	1	1	0	1	0	0
30	115		1	0	1	1	0	1	0	1
31	116		1	0	1	1	0	1	1	0
32	117		1	0	1	1	0	1	1	1
33	118		1	0	1	1	1	0	0	0
34	119		1	0	1	1	1	0	0	1
35	120		1	1	0	0	0	0	0	0
36	121		1	1	0	0	0	0	0	1
37	122		1	1	0	0	0	0	1	0
38	123		1	1	0	0	0	0	1	1
39	124		1	1	0	0	0	1	0	0
40	125		1	1	0	0	0	1	0	1

"1" = Switch Open
"2' = Switch Closed

tors, so an ungrounded or open pin will be at the binary 1 level.

Notice that the output of the CMOS oscillator/buffer is connected directly to the input of the divide-by-512 reference frequency divider. The reference frequency for the frequency/phase detector is 10 kHz. The reference oscillator operates at 5.12 MHz.

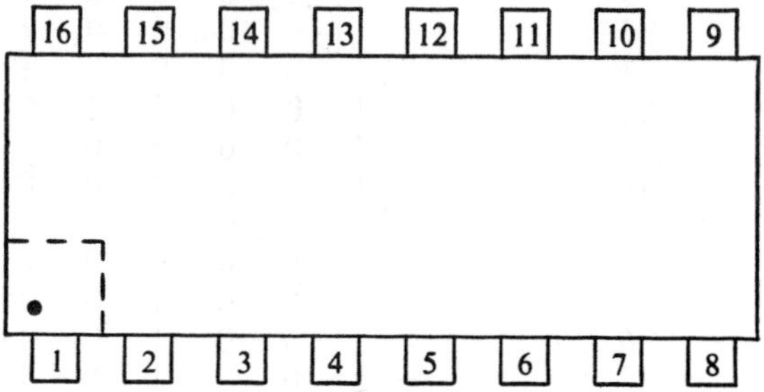

Fig. 3-3. An overview of the SC42502P/3001-201 IC, showing the pin numbering scheme. Note that the number 1 pin is nearest the dot.

A TYPICAL APPLICATION

The block diagram in Fig. 3-4 shows a typical application of the SC42502P/3001-201 IC. Using channel 40 as the example, here is how the circuit operates. The VCO operates in the 22 to 23 MHz range. For channel 40, the programmable divider is set to divide its input by 125 (see Table 3-1). In order for the output of the divider to be 10 kHz, the input to the programmable divider will have to be 125 × 10 kHz, or 1.250 MHz. A signal is fed from the VCO to the synthesis mixer where it is mixed with the 21.855-MHz signal from the high oscillator. The difference of these two frequencies is sent to the programmable divider input at pin 1 of the SC42502P/3001-201 IC. On channel 40, the input to the programmable divider has to be 1.25 MHz. This means that the VCO frequency has to be 1.25 MHz + 21.855 MHz, or 23.105 MHz. If the VCO frequency is too high or too low, the output of the programmable divider will then be higher or lower than the 10-kHz reference frequency. This will cause the phase detector to develop a correction voltage. This correction voltage is applied to the control line of the VCO at pin 8 of the IC. Notice the RC filter network connected in the control line. The network consists of R1, R2, R3, C1, and C2. This filter network smooths out the pulses from the phase detector. The output from the filter network is a

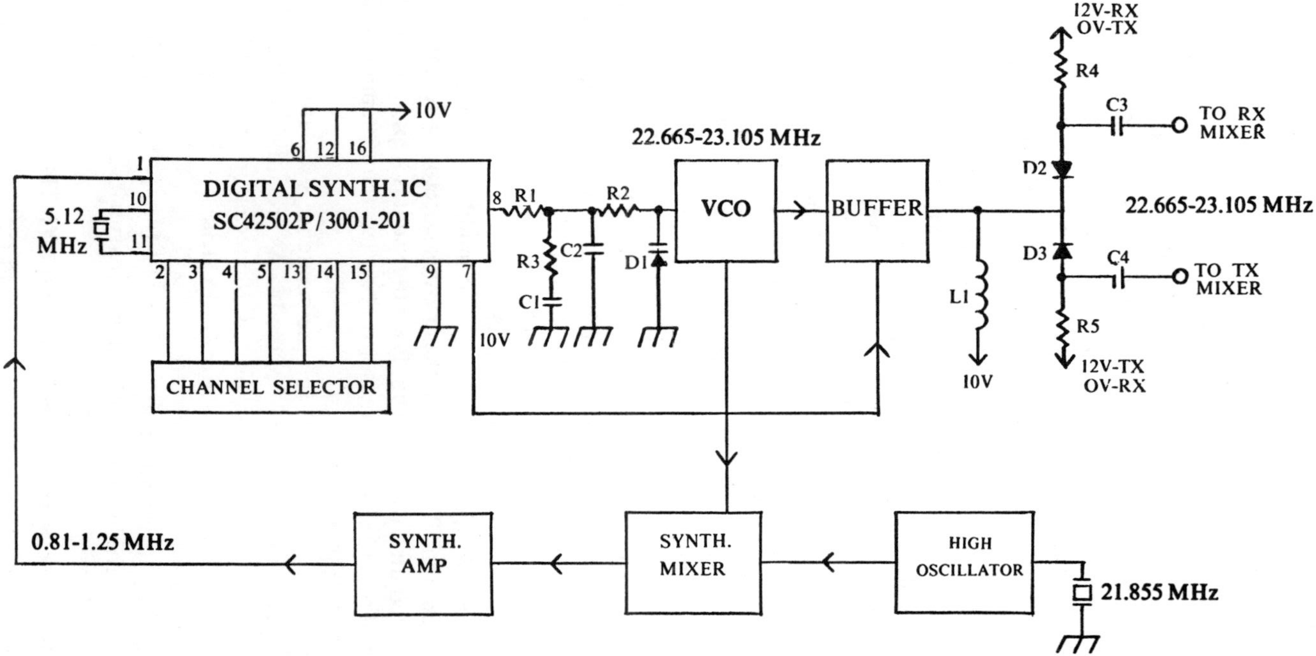

Fig. 3-4. A block diagram of a typical application of the SC42502P/3001-201 IC. This PLL requires three crystals. Only two are shown; the other is at the transmit oscillator, which operates at 4.3 MHz. This very same arrangement is used in some Johnson models.

smooth DC control voltage. This control voltage causes the VCO to "search" until the output of the programmable divider is exactly 10 kHz. This will occur when the VCO frequency is 23.105 MHz. This VCO frequency is sent to the buffer. From the buffer, the VCO signal is applied to the cathode of diodes D2 and D3. These diodes serve to "steer" the VCO signal to the correct circuit for the receive and transmit modes of operation. In the receive mode, diode D2 is forward biased, while diode D3 is reversed biased. Thus, in the receive mode the VCO signal will pass through D2 and then through C3 to the receiver mixer. In the receiver mixer, the 21.105-MHz VCO signal is mixed with the incoming 27.405-MHz signal (channel 40) to yield a difference frequency of 27.405 MHz - 23.105 MHz, or 4.3 MHz. This 4.3-MHz signal is the receiver intermediate frequency.

In the transmit mode, diode D2 is reversed biased, while diode D3 is forward biased. In the transmit mode, therefore, the VCO signal will pass through diode D3 and then through C4 on way to the transmit mixer. At the transmit mixer, the VCO signal is mixed with a 4.3 MHz signal from the transmit oscillator to yield a sum frequency of 23.105 MHz + 4.3 MHz, or 27.405 MHz. Thus, the channel 40 transmit frequency is developed.

Notice on the block diagram in Fig. 3-4 that pin 7 of the IC connects to the buffer stage. Pin 7 is HIGH (10 volts) when the loop is locked. This provides bias voltage to the buffer stage. If the loop is not locked, pin 7 goes LOW. This kills the bias to the buffer stage and thus disables the stage. This action prevents the radio from radiating improper frequencies.

Notice also that all of the programmable divider program lines, except pin 12, connect to the channel selector switch. Pin 12 is connected directly to the supply voltage so it is kept at binary 1 level at all times. Since the divisor is never less than 80, pin 12 will never need to be low.

The 5.12-MHz crystal for the reference oscillator is connected between pins 10 and 11 of the IC. As previously noted, the output of the reference oscillator is connected internally to the divide-by-512 fixed reference divider to yield a 10-kHz reference for the frequency/phase detector.

One slightly unusual feature of this PLL IC is the large difference between the binary 1 level and the binary 0 level. Binary 1 = 10 volts, and binary 0 = 0 volts.

E. F. JOHNSON MESSENGER 4140 SYNTHESIZER

The E. F. Johnson Messenger 4140 CB radio is pictured in

Fig. 3-5. the block diagram of the synthesizer circuit is shown in Fig. 3-6, and the complete schematic of the synthesizer is shown in Fig. 3-7. Refer to both Figs. 3-6 and 3-7 during this discussion.

The reference oscillator operates at 5.12 MHz. Note in Fig. 3-7 that the 5.12-MHz reference oscillator crystal (Y101) connects between pins 10 and 11 of the digital synthesizer IC (U101). Pin 11 is the input to a CMOS oscillator, while pin 10 is the output of the oscillator. Note from Fig. 3-7 that a portion of the 5.12-MHz signal is fed to the offset mixer stage at Q102.

The VCO stage consists of transistor Q101 and the associated components. This VCO operates in the 31-32 MHz range. The chart in Table 3-2 lists the VCO frequencies for each of the 40 channels for both the receive and the transmit modes. In some PLL synthesizers, the VCO frequency is the same for the receive and the transmit modes, but in this synthesizer the VCO operates at different frequencies for the receive and transmit modes. Notice from the chart in Table 3-2 that on any given channel, the VCO frequency for the transmit mode is 455 kHz higher than for the receive mode. For the receive mode, the VCO frequency ranges from 31.630 MHz for channel 1 to 32.070 MHz for channel 40. For the transmit mode, the VCO frequency ranges from 32.085 MHz for channel 1 to 32.525 MHz for channel 40. The chart in Table 3-2 also shows the programmable divider input frequency for channels 1 through 40. This ranges from 0.810 MHz for channel 1 to 1.250 MHz for channel 40.

In Fig. 3-7, note C103, C104, C106, C107, T101, and varactor diode CR101. All of these components form the *resonant circuit* for the VCO. Coil T101 is used to set the VCO in the proper frequency range. Then the control voltage from pin 8 of U101 fine tunes the VCO by controlling the reverse bias on the varactor diode, CR101.

The offset mixer (Q102) mixes the VCO frequency with the 5.120-MHz signal from the reference oscillator. A difference frequency, (FVCO-5.12 MHz) appears at the output of the offset mixer stage. This difference frequency will not be the same for the transmit and receive modes. In the receive mode, the signal is 455 kHz below the channel frequency. This signal is fed to the receiver mixer where it is heterodyned with the incoming receive signal to produce a difference frequency of 455 kHz signal is the receiver intermediate frequency. This receiver is of the single conversion type because only one intermediate frequency (455 kHz) is used.

In the transmit mode, the signal at the output of the offset mixer is equal to the channel frequency. This signal is sent directly to the transmitter amplifier stages where it is amplified and fed to the antenna.

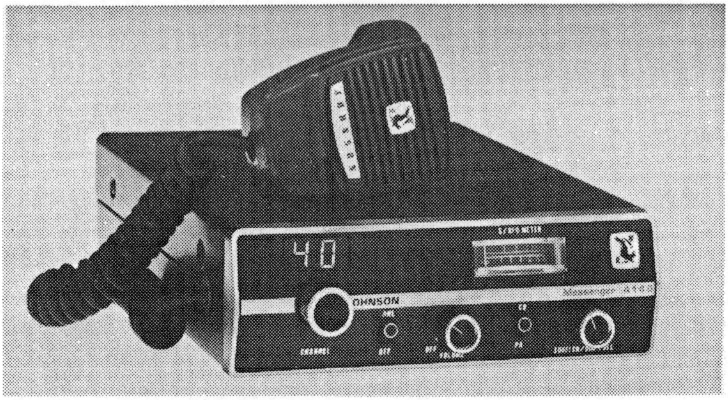

Fig. 3-5. The E. F. Johnson Messenger 4140 CB radio (courtesy of E. F. Johnson Co.).

The high oscillator consists of transistor Q104, crystals Y102 and Y103, and the associated components. Notice that each crystal connects to ground through a diode. In the receive mode, diode CR104 is biased on. This connects Y102 in the circuit. At the same time diode, CR105 is cut off because of lack of forward bias. Thus Y103 is effectively out of the circuit. In the transmit mode, just the opposite situation exists. Diode CR104 is cut off because of lack of forward bias, so now crystal Y102 is switched out, while crystal Y103 is switched in. The high oscillator must operate at different frequencies for the transmit and receive modes because the VCO operates at different frequencies for the two modes. The difference between the high oscillator and the VCO frequencies remains the same for the transmit and receive modes because both the high oscillator frequency and the VCO frequency shifts the same amount between the transmit and receive modes. This keeps the input to the programmable divider constant for the transmit and receive modes on a given channel. The VCO frequency is mixed with the high oscillator frequency in the synthesizer mixer, Q105. The difference frequency is amplified by transistor Q106 and fed to the input of the programmable divider.

The following is a complete analysis of the PLL operation using channel 40 as the example. First, let's look at the PLL operation in the receive mode. When the CB set is first switched on, the VCO will operate at some random frequency in the 31-32 MHz range. Assume that when power is applied, the VCO initially

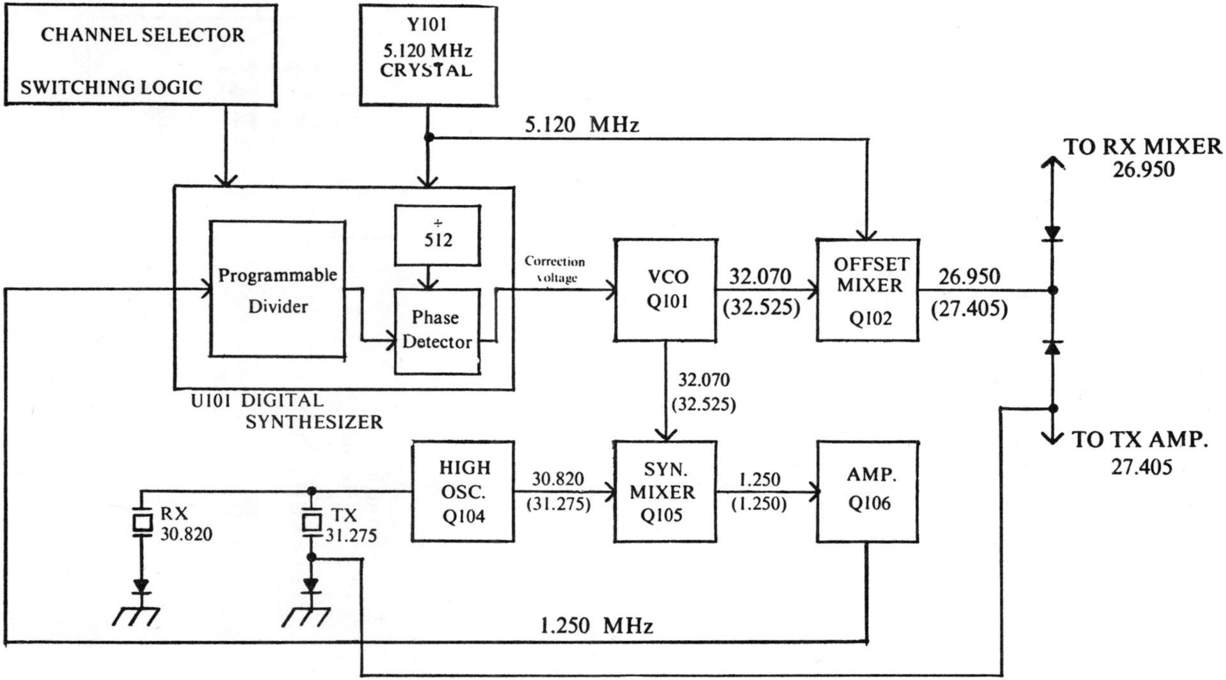

Fig. 3-6. Block diagram of the Messenger 4140 digital synthesizer circuit. The frequencies shown are for channel 40. The frequency shown in parentheses is for transmit mode; the other is for the receive mode (courtesy of E. F. Johnson Co.).

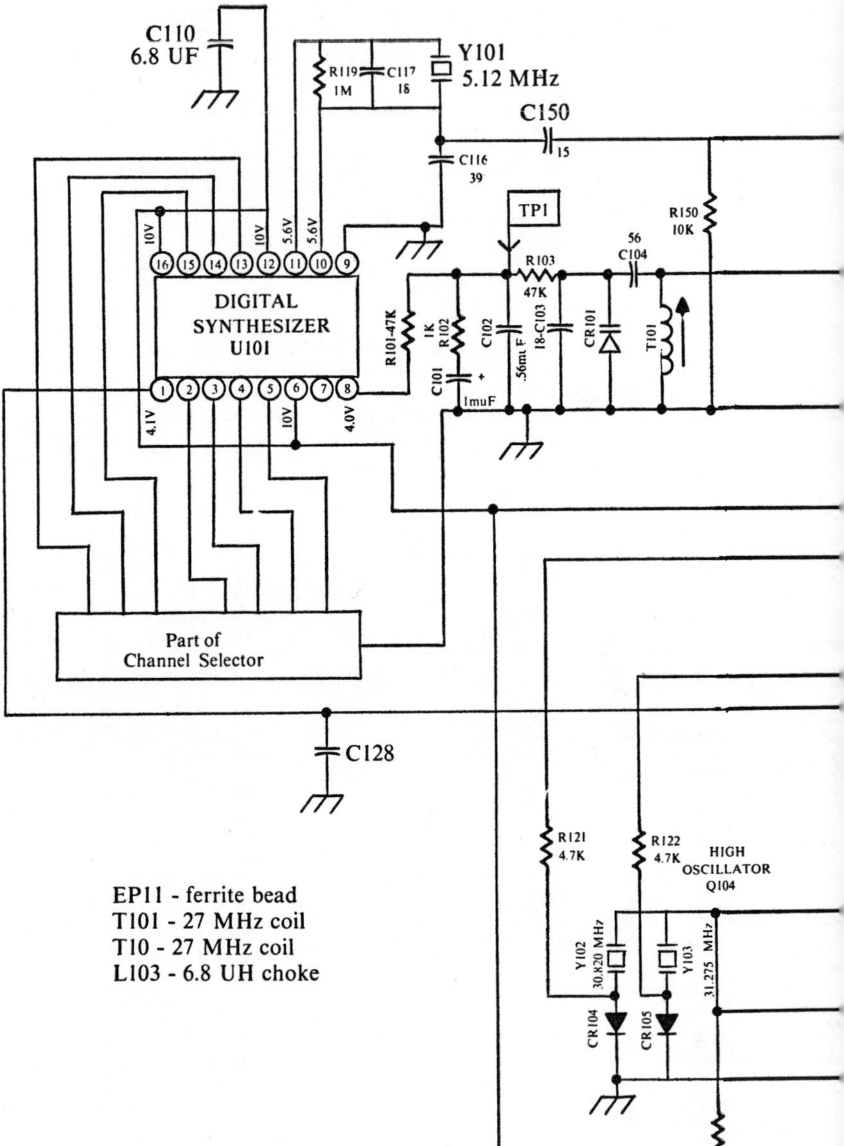

Fig. 3-7. Schematic diagram of the Messenger 4140 digital synthesizer circuit.

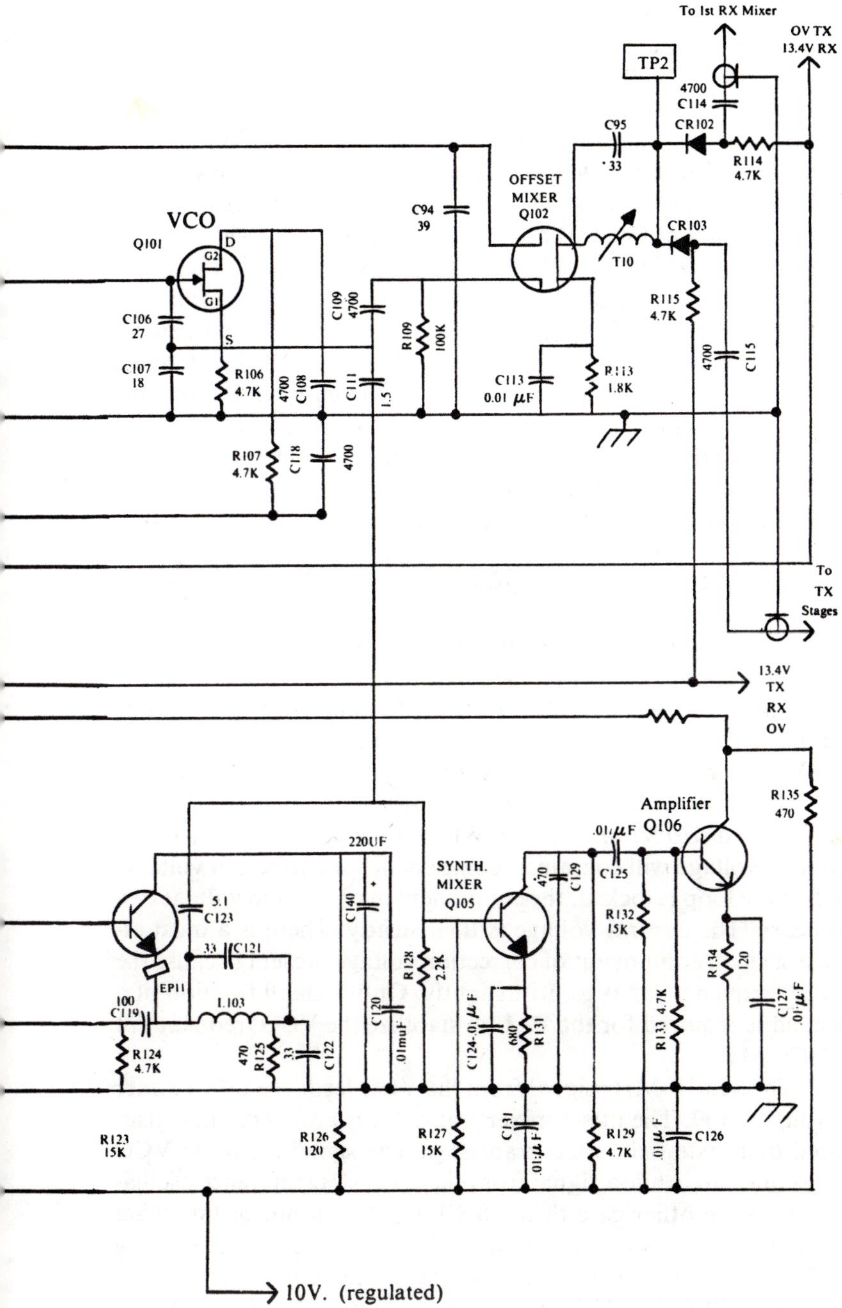

To 1st RX Mixer
OV TX
13.4V RX
TP2
4700
C114
C95
CR102
.33
R114
4.7K
OFFSET
MIXER
Q102
VCO
Q101
D
G2
G1
S
C94
39
CR103
T10
C106
27
C107
18
C109
4700
R109
100K
R115
4.7K
R106
4.7K
4700
C108
C111
1.5
C113
0.01 μF
R113
1.8K
4700
C115
R107
4.7K
C118
4700
To
TX
Stages
13.4V
TX
RX
OV
R135
470
Amplifier
Q106
.01μF
C125
220UF
C140
SYNTH.
MIXER
Q105
470
C129
5.1
C123
R132
15K
33
C121
EP11
R128
2.2K
R134
120
100
C119
L103
C120
.01muF
C124-.01μF
680
R131
R133 4.7K
C127
.01μF
R124
4.7K
470
R125
33
C122
R123
15K
R126
120
R127
15K
C131
.01μF
R129
4.7K
.01μF
C126
10V. (regulated)

starts operating at 31.500 MHz. A sample of this 31.500-MHz signal is fed to the synthesizer mixer where it is mixed with the signal from the high oscillator. The high oscillator frequency in the receive mode is 30.820 MHz. At the output of the synthesizer mixer, there appears the difference frequency of FVCO-30.820 MHz. At this particular instant, this difference frequency is 31.500 MHz - 30.820 MHz, or 0.680 MHz. This 0.680-MHz difference signal is amplified by Q106 and fed to the input of the programmable divider at pin 1 of U101. When set to channel 40, the divider is programmed to divide by 125. The 0.680-MHz signal is therefore divided by 125 to yield 5.44 kHz at the output of the programmable divider. This 5.44 kHz signal is then sent to the frequency/phase comparator where it is compared with the 10-kHz reference frequency. Because a large difference exists between the two inputs to the frequency/phase detector, a large error signal will be produced at the output of the frequency/phase detector at pin 8 of U101. This error signal is filtered by the RC network at pin 8 and the result is a relatively pure DC correction voltage. This correction voltage changes the bias on the varactor diode, CR101, to move the VCO frequency upward. As the VCO frequency increases, the input to the programmable divider approaches 1.250 MHz and the output of the divider approaches 10 kHz. As this occurs, the correction voltage becomes smaller, also. Finally, when the VCO reaches 32.070 MHz, the input to the programmable divider output will be 10 kHz and the correction voltage to the VCO control line will be 0. A control voltage will be applied to the VCO control line, but there is a distinction between *control voltage* and *correction voltage.* When the loop is not locked, the control voltage will contain a "component" of correction voltage. When the loop is locked, the component of correction voltage disappears. The control voltage is then steady. There is almost always a small component of correction voltage present because the VCO frequency tends to drift slightly. Only a small fraction of a second is required for the PLL to stabilize the VCO frequency at 32.070 MHz.

The 32.070-MHz signal from the VCO feeds the offset mixer through C109. The offset mixer transistor is an N-channel, insulated dual-gate field effect transistor. The signal from the VCO feeds one gate while a signal from the 5.12-MHz reference oscillator feeds the other gate through C150. The output of the offset mixer is taken from the drain through the tuned circuit consisting of C95 and T10. This tuned circuit is resonant at the difference frequency of FVCO - 5.12 MHz, which on channel 40 is 32.070 MHz-

Table 3-2. VCO and Programmable Divider Input Frequencies.

CHANNEL NO.	VCO(TX)	VCO(RX)	Divider Input (TX & RX)
1	32.085 MHz	31.630 MHz	0.810 MHz
2	32.095 "	31.640 "	0.820 "
3	32.105 "	31.650 "	0.830 "
4	32.125 "	31.670 "	0.850 "
5	32.135 "	31.680 "	0.860 "
6	32.145 "	31.690 "	0.870 "
7	32.155 "	31.700 "	0.880 "
8	32.175 "	31.720 "	0.900 "
9	32.185 "	31.730 "	0.910 "
10	32.195 "	31.740 "	0.920 "
11	32.205 "	31.750 "	0.930 "
12	32.225 "	31.770 "	0.950 "
13	32.235 "	31.780 "	0.960 "
14	32.245 "	31.790 "	0.970 "
15	32.255 "	31.800 "	0.980 "
16	32.275 "	31.820 "	1.000 "
17	32.285 "	31.830 "	1.010 "
18	32.295 "	31.840 "	1.020 "
19	32.305 "	31.850 "	1.030 "
20	32.325 "	31.870 "	1.050 "
21	32.335 "	31.880 "	1.060 "
22	32.345 "	31.890 "	1.070 "
23	32.375 "	31.920 "	1.100 "
24	32.355 "	31.900 "	1.080 "
25	32.365 "	31.910 "	1.090 "
26	32.385 "	31.930 "	1.110 "
27	32.395 "	31.940 "	1.120 "
28	32.405 "	31.950 "	1.130 "
29	32.415 "	31.960 "	1.140 "
30	32.425 "	31.970 "	1.150 "
31	32.435 "	31.980 "	1.160 "
32	32.445 "	31.990 "	1.170 "
33	32.455 "	32.000 "	1.180 "
34	32.465 "	32.010 "	1.190 "
35	32.475 "	32.020 "	1.200 "
36	32.485 "	32.030 "	1.210 "
37	32.495 "	32.040 "	1.220 "
38	32.505 "	32.050 "	1.230 "
39	32.515 "	32.060 "	1.240 "
40	32.525 "	32.070 "	1.250 "

5.12 MHz, or 26.950 MHz. This is applied to the junction of diodes CR102 and CR103. In the receive mode, only CR102 is forward biased, so the signal is routed through CR102 on to the receiver mixer where it is heterodyned with the incoming receive signal to producc a difference frequency that is the intermediate frequency. On channel 40 the channel frequency is 27.405 MHz. This is mixed with the 26.950-MHz signal from the offset mixer to produce a difference frequency of 27.405 MHz - 26.950 MHz, or 455 kHz. This is the only intermediate frequency used in this set.

Now examine the transmit mode of operation. When the push-to-talk switch is actuated, diode CR105 in the high oscillator circuit will become forward biased while diode Cr104 loses its forward bias. This disconnects Y102 (the 30.820-MHz xtal) from the oscillator and connects Y103 (the 31.275-MHz xtal) to the oscillator. So the high oscillator is shifted to a new frequency, 31.275 MHz. At the instant the transmitter is keyed, the VCO will still be operating at 32.070 MHz. So momentarily, the input to the programmable divider will be 32.070 MHz - 31.275 MHz, or 0.795 MHz. The output from the programmable divider will be 0.795 MHz divided by 125, or 6.36 kHz. This is fed to the frequency/phase detector where it is compared with the 10-kHz reference signal. The frequency/phase comparator detects the large frequency difference of its two inputs and issues a correction voltage that changes the bias on the varactor diode and moves the VCO frequency upward. As the VCO frequency moves upward the input to the programmable divider approaches 1.25 MHz and the output of the programmable divider approaches the 10-kHz reference frequency. As this occurs, the correction signal approaches O. When the VCO frequency reaches 32.525 MHz, the input to the programmable divider will be 1.25 MHz, the output of the programmable divider will be 10-kHz, and the control voltage to the VCO will no longer contain an error component. The control voltage now stabilizes the VCO frequency at 32.525 MHz. As in the receive mode, the VCO frequency is mixed with the 5.12-MHz oscillator frequency in the offset mixer to produce a difference frequency at the output (drain). The difference frequency at the output of the offset mixer is 32.525 MHz - 5.12 MHz, or 27.405 MHz. This is applied to the junction of diodes CR102 and CR103. In the transmit mode, only CR103 is forward biased, so the signal is routed to the transmitter amplifier stages.

This completes the basic analysis of this PLL circuit. Remember that when the frequency of the high oscillator is shifted (as when going between the transmit and receive modes), the VCO must be shifted also in order to keep the input to the input to the phase detector (from the programmable divider) at 10-kHz.

Alignment

☐ Using a frequency counter, check at pin 10 of U101 for 5.12 MHz. The frequency there should be within 200 to 300 Hz of 5.12 MHz. There is no adjustment provided to *zero* the 5.12 MHz oscillator on frequency.

☐ Check the frequency of the high oscillator in the *receive* mode at the emitter of transistor Q104. The frequency should be within 200 to 300 Hz of 30.820 MHz.

☐ Check the frequency of the high oscillator in the *transmit* mode at the emitter of transistor Q104. The frequency should be within 200 to 300 Hz of 31.275 MHz.

☐ With the channel selector switched to channel 20, connect a DC voltmeter to TP1 and adjust T101 for 4.0 volts with the core of T101 at the peak closest to the PC board.

☐ With the channel selector switched to channel 20, adjust T10 for maximum rf voltage at TP2. Use an rf voltmeter or wideband oscilloscope.

☐ Using a frequency counter, check for the proper frequencies at TP2 for all channels on transmit and receive modes. Refer to Table 3-3 for the proper frequencies.

Troubleshooting

When the set neither transmits nor receives, check the supply voltage to the PLL synthesizer. This can be checked at pin 16 of U101. If the supply voltage is present, check for rf voltage at TP2. If there is rf voltage at TP2, check diodes CR102 and CR103. Also check to see that they are properly biased for the transmit modes. If there is no rf voltage at TP2, check for rf voltage on the two gates of Q102. On gate 2, there should be an rf voltage at a frequency of 5.12 MHz. On gate 1, there should be an rf voltage from the VCO. For the specific frequency, check Table 3-2 for the specific channel and mode of operation. If the 5.12-MHz signal is missing on gate 2, use an rf voltmeter and trace back to the 5.12-MHz oscillator at pin 10 of U101. Troubleshoot the reference oscillator if there is no rf voltage at pin 10 of U101. If the VCO signal is missing from the gate 1 of Q102, troubleshoot the VCO. If proper signals are present on gates 1 and 2, troubleshoot the offset mixer stage.

In this example, it was assumed that there was no rf output from the PLL synthesizer to either the transmitter or receiver

Table 3-3. Transmit and Receive Mode Frequencies at Test Point 2.

Channel No.	Frequency At TP2 RX MODE (MHz)	Frequency At TP2 TX MODE (MHz)
1	26.510	26.965
2	26.520	26.975
3	26.530	26.985
4	26.550	27.005
5	26.560	27.015
6	26.570	27.025
7	26.580	27.035
8	26.600	27.055
9	26.610	27.065
10	26.620	27.075
11	26.630	27.085
12	26.650	27.105
13	26.660	27.115
14	26.670	27.125
15	26.680	27.135
16	26.700	27.155
17	26.710	27.165
18	26.720	27.175
19	26.730	27.185
20	26.750	27.205
21	26.760	27.215
22	26.770	27.225
23	26.800	27.255
24	26.780	27.235
25	26.790	27.245
26	26.810	27.265
27	26.820	27.275
28	26.830	27.285
29	26.840	27.295
30	26.850	27.305
31	26.860	27.315
32	26.870	27.325
33	26.880	27.335
34	26.890	27.345
35	26.900	27.355
36	26.910	27.365
37	26.920	27.375
38	26.930	27.385
39	26.940	27.395
40	26.950	27.405

Table 3-4. Various Segments that are Energized for Display.

CHANNEL NO.	TENS CHANNEL INDICATOR UNITS													
	A	B	C	D	E	F	G	A	B	C	D	E	F	G
1									X	X				
2								X	X		X	X		X
3								X	X	X	X			X
4									X	X			X	X
5								X		X	X		X	X
6										X	X	X	X	X
7								X	X	X				
8								X	X	X	X	X	X	X
9								X	X	X			X	X
10		X	X					X	X	X	X	X	X	
11		X	X						X	X				
12		X	X					X	X		X	X		X
13		X	X					X	X	X	X			X
14		X	X						X	X			X	X
15		X	X					X		X	X		X	X
16		X	X							X	X	X	X	X
17		X	X					X	X	X				
18		X	X					X	X	X	X	X	X	X
19		X	X					X	X	X			X	X
20	X	X		X	X		X	X	X	X	X	X	X	
21	X	X		X	X		X		X	X				
22	X	X		X	X		X	X	X		X	X		X
23	X	X		X	X		X	X	X	X	X			X
24	X	X		X	X		X		X	X			X	X
25	X	X		X	X		X	X		X	X		X	X
26	X	X		X	X		X			X	X	X	X	X
27	X	X		X	X		X	X	X	X				
28	X	X		X	X		X	X	X	X	X	X	X	X
29	X	X		X	X		X	X	X	X			X	X
30	X	X	X	X			X	X	X	X	X	X	X	
31	X	X	X	X			X		X	X				
32	X	X	X	X			X	X	X		X	X		X
33	X	X	X	X			X	X	X	X	X			X
34	X	X	X	X			X		X	X			X	X
35	X	X	X	X			X	X		X	X		X	X
36	X	X	X	X			X			X	X	X	X	X
37	X	X	X	X			X	X	X	X				
38	X	X	X	X			X	X	X	X	X	X	X	X
39	X	X	X	X			X	X	X	X			X	X
40		X	X			X	X	X	X	X	X	X	X	

stages. Note that this synthesizer does not use the lock detector output to disable the transmitter when an out-of-lock condition occurs. So it is possible for the transmitter to radiate, even though the PLL synthesizer may be out-of-lock.

Wrong Transmit and/or Receive Frequencies on High Channels with Low channels OK. Check the VCO alignment. Before adjusting the VCO, first check the high oscillator frequency for accuracy. Of course, if the high oscillator were operating off frequency in either mode, the VCO frequency would be wrong on all channels. Still, before adjusting the VCO, check the high oscillator frequency. See the "Alignment" section for proper adjustment of the VCO. If the high oscillator is operating normally, abnormal voltage at pin 8 of U101 is a clue that the VCO needs adjustment.

Wrong Transmit and/or Receive Frequencies on Low Channels with High Channels OK. This is about the same trouble as just mentioned except this time the VCO is probably out of adjustment in the *opposite* direction. See the "Alignment" section for proper VCO adjustment. Again, be sure to first check the high oscillator frequency before adjusting the VCO. Check for adnormal voltage at pin 8 of U101.

Weak Receive or No Receive, Off-Frequency Transmit with All Channels. In this case the synthesizer output frequency is wrong for both the receive and transmit modes on all channels. If only one mode were affected, one of the high-oscillator crystals would be a suspect. But it is unlikely that both crystals would fail simultaneously. Even so, it is still possible, so this can't be neglected. Also, possibly another component failure in the high oscillator could shift the frequency. Switch diodes CR104 and CR105, if necessary.

If the high oscillator is OK, check the 5.12-MHz oscillator frequency. If this frequency is wrong, the transmit and receive frequencies will all be wrong.

Transmit Frequencies Wrong with Receive OK. Check the high oscillator transmit crystal. Also check for proper bias on the switching diode, CR105. If switching diode CR104 were shorted, both crystals would be connected to the high oscillator in the transmit mode. This would cause improper operation.

Receive Frequencies Wrong with Transmit OK. Check the high-oscillator receive crystal. Also check diode CR104, the bias to CR104, and CR105.

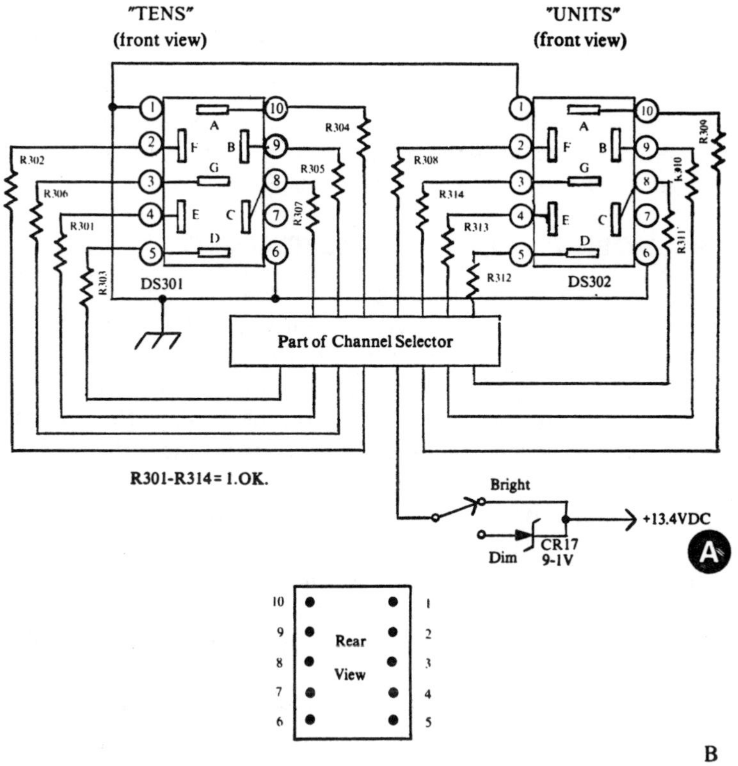

Fig. 3-8. The wiring of the two seven-segment displays used for channel indication is shown at A, while a rear view of the seven-segment display, showing the pin numbering scheme, is shown at B.

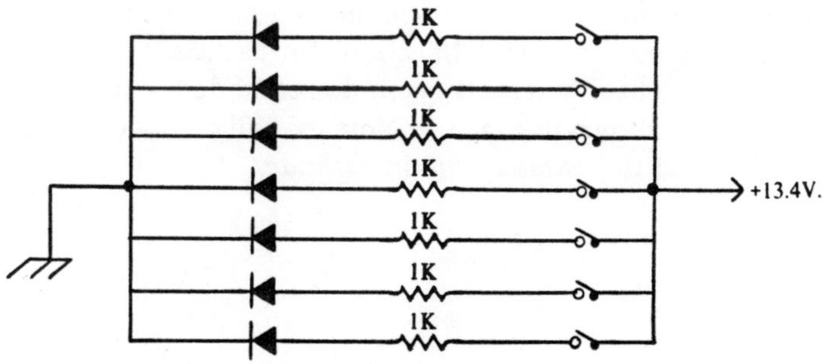

Fig. 3-9. An equivalent circuit of the seven-segment LED display. Note that all cathodes are connected for a common cathode arrangement.

Some Channels Inoperative. Check the channel selector switch. Check the programming lines to the digital synthesizer IC, U101, for the proper binary levels. A defective U101 could be the trouble.

Some Channels Intermittent. This is probably caused by dirty or loose contacts on the channel selector switch. Try cleaning the selector switch with a good tuner cleaner.

MESSENGER 4140 LED DIGITAL READOUT

Figure 3-8A shows the diagram of the digital readout LED channel indicator. There are two, seven-segment LED displays used, one for the "tens" and the other for the "units" digit. This display is of the common cathode type. The cathodes of all segments are tied together and connected to pin 6 of the display. The anodes are brought out to separate pins of the display. The cathodes are all grounded, and each anode connects to the channel selector switch through a 1000-ohm, series current-limiting resistor. The various segments are rated at 1.63 volts 23 mA. The current-limiting resistors hold the current to less than 14 mA. Notice that the channel selector switch common terminal connects to the supply voltage through a switch. In the *bright* position, the full supply voltage is fed to the switch. In the *dim* position, the supply voltage is connected through CR17 to the channel switch. CR17 is a 9.1-volt zener diode, so in this position the voltage to the channel selector is reduced by the zener diode. The *dim* position is much more comfortable to view in the darkness.

The channel selector switch supplies voltage to the proper segments for each channel. Figure 3-8B shows the pin numbering scheme for the LED display. The chart in Table 3-4 shows which segments are energized for each channel number, 1 through 40. Figure 3-8A shows the pin connections for each segment.

An equivalent circuit for the common cathode seven-segment LED display is shown in Fig. 3-9. Note that all cathodes are tied together, hence, the name *common cathode.*

Chapter 4
The REC86345 Digital Synthesizer IC

The REC86345 digital synthesizer IC is widely used. This IC was developed by Resdel Engineering Corporation, the parent company of Fanon/Courier. This synthesizer IC features a sample-and-hold phase detector which provides exceptional low phase noise output of the VCO without using active filters.

SPECIAL FEATURES

- ☐ Low Phase Noise Sample-and-Hold Phase Detector
- ☐ Lock-Detector Output (clean HIGH or LOW)
- ☐ Low Power, Typically Less Than 10 mW Dissipation
- ☐ On-Chip Pulldown Resistors
- ☐ On-Chip Reference Oscillator (with external crystal)
- ☐ Selectable Phase Detector Reference Frequency (5 kHz or 10 kHz)
- ☐ Eight Binary Inputs to Programmable Divider

INTERNAL STRUCTURE

The block diagram in Fig. 4-1 shows the internal structure of the REC86345 IC. Figure 4-2 shows the pin numbering scheme and pin functions of the IC.

Reference Section

This section consists of a CMOS inverter that can be used as

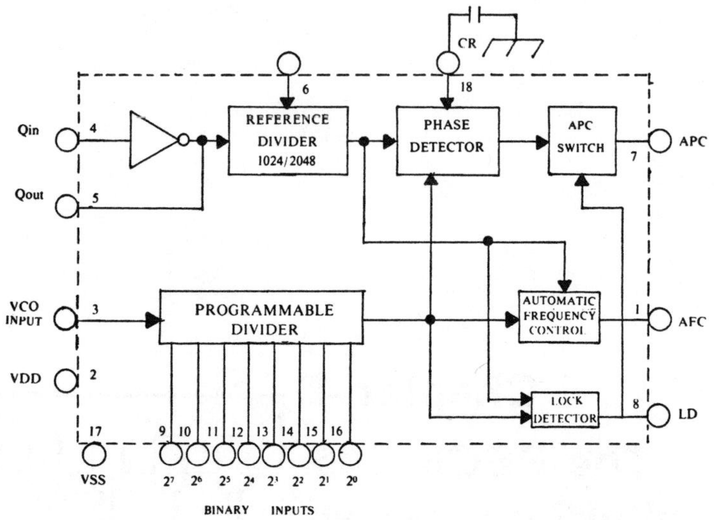

Fig. 4-1. The internal structure of the Resdel REC86345 PLL IC.

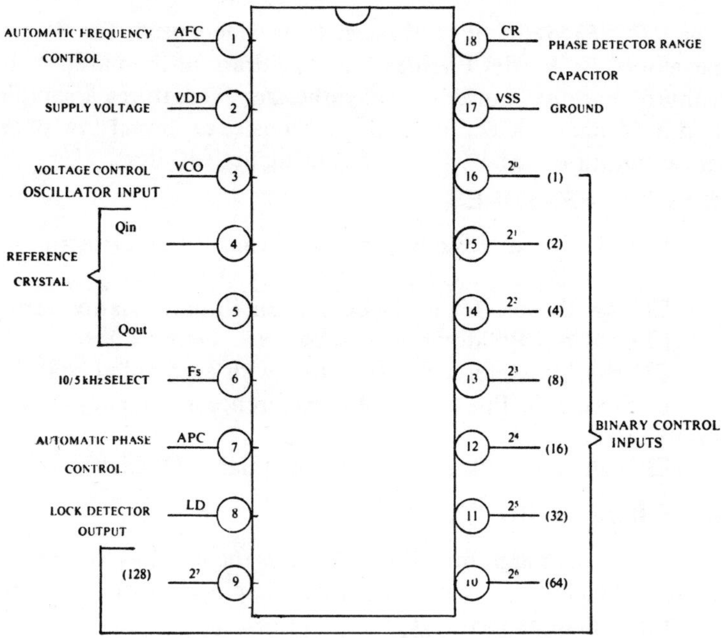

Fig. 4-2. An overview of the Resdel REC86345 PLL IC showing the functions of the various pins and the pin numbering scheme.

the reference oscillator by connecting the proper crystal between the input and the output of the CMOS inverter at pins 4 and 5 of the IC. The reference divider can be programmed to divide by either 1024 or 2048. This makes it possible to have two different reference frequencies available, 5 kHz or 10 kHz, depending upon the particular application. The reference frequency is determined by the voltage level at pin 6 as follows: If pin 6 is open, the reference divider divides by 1024 to yield a 10-kHz reference frequency. If pin 6 is high, the divider divides by 2048 to yield a 5-kHz reference frequency.

Phase Detector

This phase detector is of the sample and hold design with an internal hold capacitor of approximately 30 pF. The ramp forming capacitor, CR, is charged via an internal current generator of approximately 90 □ 20 microamps.

Automatic Phase Control Switch

The automatic phase control switch connects the APC output to the VCO control line when the frequency detector indicates a frequency acquisition.

Programmable Divider

This divider has eight binary input control lines. The maximum divide by N number possible for this divider is 255. Each program input has an internal pulldown resistor which is built on the IC chip itself. The pulldown current is approximately 100 microamps. The programmable divider can properly divide input frequencies up to 2.5 MHz.

The programming is applied to inputs pins 9, 10, 11, 12, 13, 14, 15, and 16. The programming is done in the pure binary form. The weights of the programming pins are listed in Table 4-1. To determine what divisor (N) is programmed into the programmable divider, simply add the weights of the pins where the binary 1 level appears. Binary 1 level is represented by approximately 4.7 volts, and binary 0 level is represented by 0 or near 0 volts.

Automatic Frequency Control

This circuit brings the VCO frequency to within the lock range of the phase detector. The automatic frequency control output (at pin 1) is a tri-state output that is open when the circuit is in a phase-locked condition, with positive-going pulses when the VCO frequency is too low and negative-going pulses when the VCO frequency is too high.

Table 4-1. Weights of Program Pins of the Resdel REC86345 PLL.

PIN NO.	"WEIGHT"
9	128 (2^7)
10	64 (2^6)
11	32 (2^5)
12	16 (2^4)
13	8 (2^3)
14	4 (2^2)
15	2 (2^1)
16	1 (2^0)

Lock Detector

This circuit provides a HIGH level (VDD) when the synthesizer attains a locked condition or a LOW level (VSS) when not in lock. The output is either HIGH or LOW. No phase pulses are developed that require filtering or detection. The lock detector output is brought out at pin 8 of the IC. This output can be used to inhibit transmitter operation if the programmed frequency cannot be properly acquire. The lock detector output will go low if a frequency error exists for more than 0.5 milliseconds.

Table 4-2 lists the absolute maximum ratings of the REC86-345 IC. Notice that the device has the capability of operating under wide temperature variations. Also notice of the soldering time and the soldering temperature. This is of particular importance to the service technician. It is important to use only enough heat to allow proper soldering of the pin and get the soldering gun or iron off the pin as quickly as possible. The chart in Table 4-3 lists the recommended operating conditions of the IC.

Table 4-2. Absolute Maximum Ratings and Recommended Operating Conditions of the Resdel REC86345 PLL.

DESCRIPTION	MINIMUM	MAXIMUM	UNIT
VDD-VSS	-0.3	+6.5	Volts
Voltage at any pin	VSS-0.3	VDD+0.3	Volts
Storage temperature	-40	+85	Degrees C
Operating temperature	-30	+70	Degrees C
Power dissipation		20	Milliwatts
Soldering temperature		230	Degrees C
Soldering time		5	Seconds

FANFARE 190 DF PLL SYNTHESIZER

All of the Fanon/Courier CB sets that use the REC86345 IC) use the same PLL circuit. A photo of the Fanfare 190 DF CB radio is shown in Fig. 4-3. The channel switch and the LED channel display are located inside the microphone assembly.

Fig. 4-3. Fanon Fanfare 190 DF CB set which uses the Resdel REC86345 PLL IC in its PLL synthesizer.

A block diagram of the PLL synthesizer circuit is shown in Fig. 4-4. The schematic diagram is shown in Fig. 4-5. Refer to Figs. 4-4 and 4-5 during the course of this discussion.

The VCO operates in the 37 to 38 MHz range. The VCO frequency is the same for the receive and transmit modes per given channel. The VCO is built around Q202. The varactor diode is D203 on the schematic. This varactor, along with capacitors C208 and C207 and transformer L201, forms the resonant circuit of the VCO. Capacitors C209, C210, and C211 also affect the resonant frequency of the VCO to some extent. The control line of the VCO connects to the cathode of varactor D203 through R208. The VCO signal is fed to the VCO buffer from the secondary of L201. The VCO signal is fed through C204 to the base of Q201 (VCO buffer transistor). The VCO buffer stage amplifies the VCO signal and then applies the signal to the receiver first mixer and to the transmit mixer.

A VCO signal is also fed from the collector of Q202 through C216 to the base of Q204. Q204 is the offset or down mixer. It mixes the VCO frequency with the 36.380-MHz signal from the offset or down oscillator to yield a difference frequency (FVCO-36.380 MHz) which feeds the input to the programmable divider. Before being applied to the input of the programmable divider,

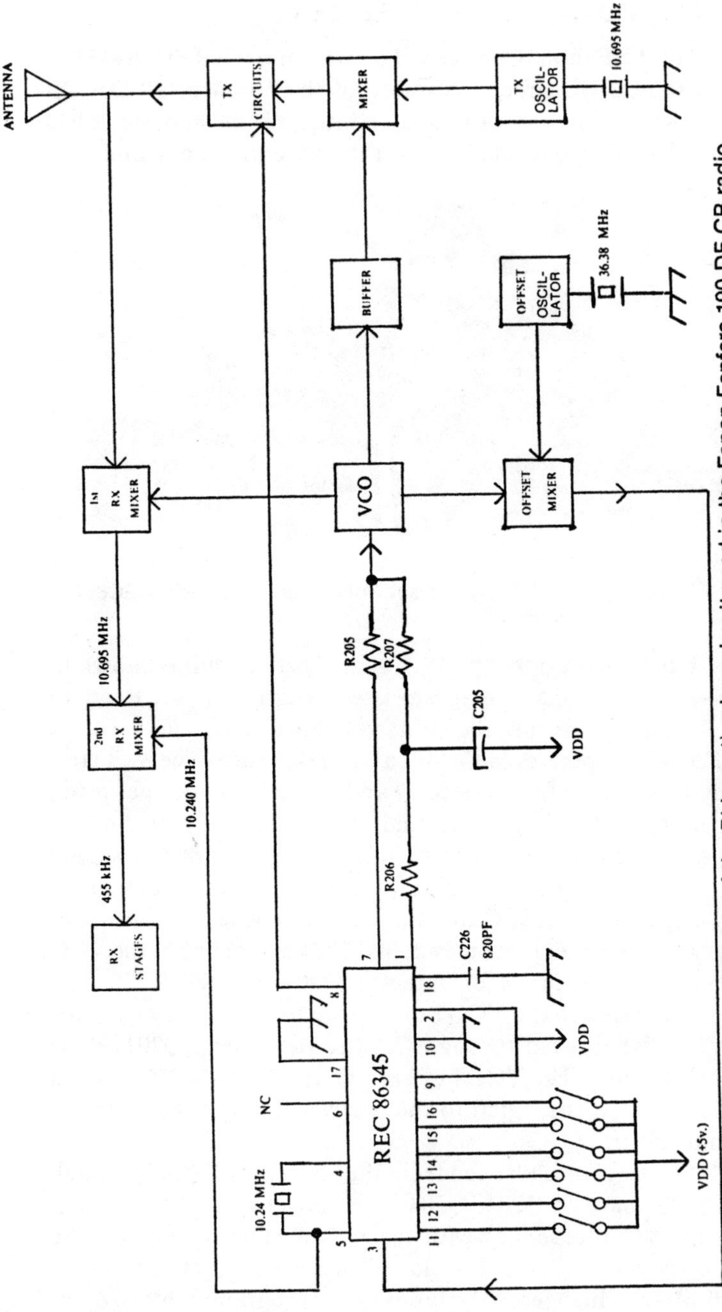

Fig. 4-4. Block diagram of the PLL synthesizer circuit used in the Fanon Fanfare 190 DF CB radio.

the difference frequency or offset frequency is amplified by the down or offset buffer stage consisting of Q205 and associated components. The offset or difference frequency is taken from the collector of Q205 and fed through C220 to the programmable divider input at pin 3 of the REC86345 IC.

The reference oscillator crystal, X202, connects between pins 4 and 5 of the IC. Pin 4 is the input to a CMOS inverter (see Fig. 4-1). The reference oscillator operates at 10.240 MHz. This 10.240-MHz signal is fed to the reference divider. This reference divider can be set to divide by 1024 or 2048, depending on the voltage level at pin 6 of the IC. If pin 6 is open, the internal pulldown resistor in the IC pulls down the pin voltage to a LOW level. In this condition the reference divider divides by 1024 to yield a 10-kHz reference frequency at the output of the reference divider. If a HIGH level voltage (+ 5 volts) is applied to pin 6 the reference divider will be set to divide by 2048. The output of the reference divider will be a 5-kHz reference frequency. Pin 6 is left open in this circuit, so the reference frequency is set at 10 kHz.

The reference oscillator signal is also buffered and fed to the transmit second mixer. Thus, the reference oscillator also serves as the second receive oscillator. The reference oscillator buffer consists of transistor Q206 and its associated components.

The programmable divider is programmed by the pure binary form. The weights of the various programming pins are shown in Fig. 4-2. To find the divisor (N) which is programmed into the divider, simply add the weights of the program pins where the binary 1 level appears. Table 4-3 shows the programming code for each of the 40 channels. Also, the input frequency to the programmable divider, the VCO frequency, and the divisor (N) are listed for each channel.

On channel 1, the programmable divider is set to divide by 128. Notice from Table 4-3 that only pin 9 is at the binary 1 level for the channel 1 position. Because pin 9 represents a weight of 128, this makes the divisor (N) 128. Notice from the table that pin 9 is always at the binary 1 level and pin 10 is always at the binary 0 level. There is no need for these two pins to connect to the channel select switch. Pin 9 is connected directly to the supply voltage to keep it at the binary 1 level at all times and pin 10 is connected directly to ground to keep it at the binary 0 level at all times. The other six programming pins (11 through 16) are connected to the channel selector switch. The channel selector either applies a HIGH level (binary 1) to the program line or it leaves the program

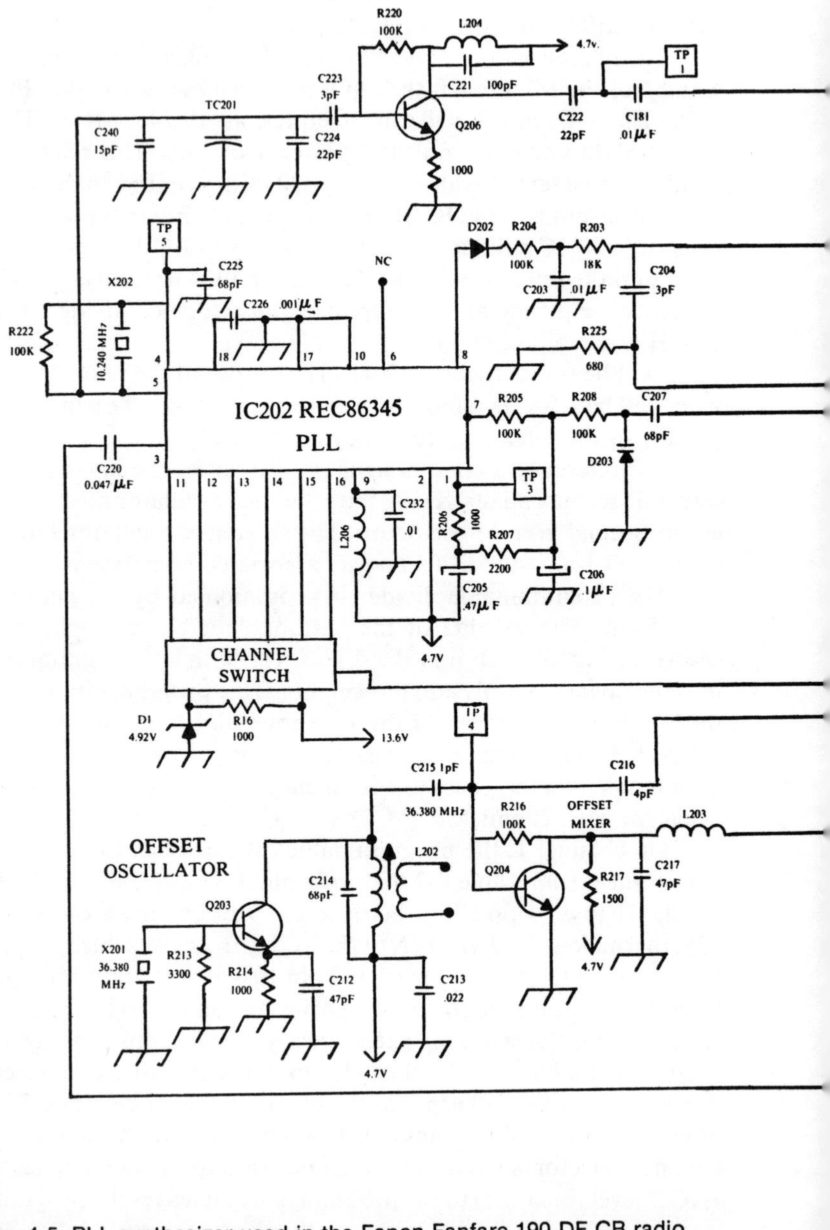

Fig. 4-5. PLL synthesizer used in the Fanon Fanfare 190 DF CB radio.

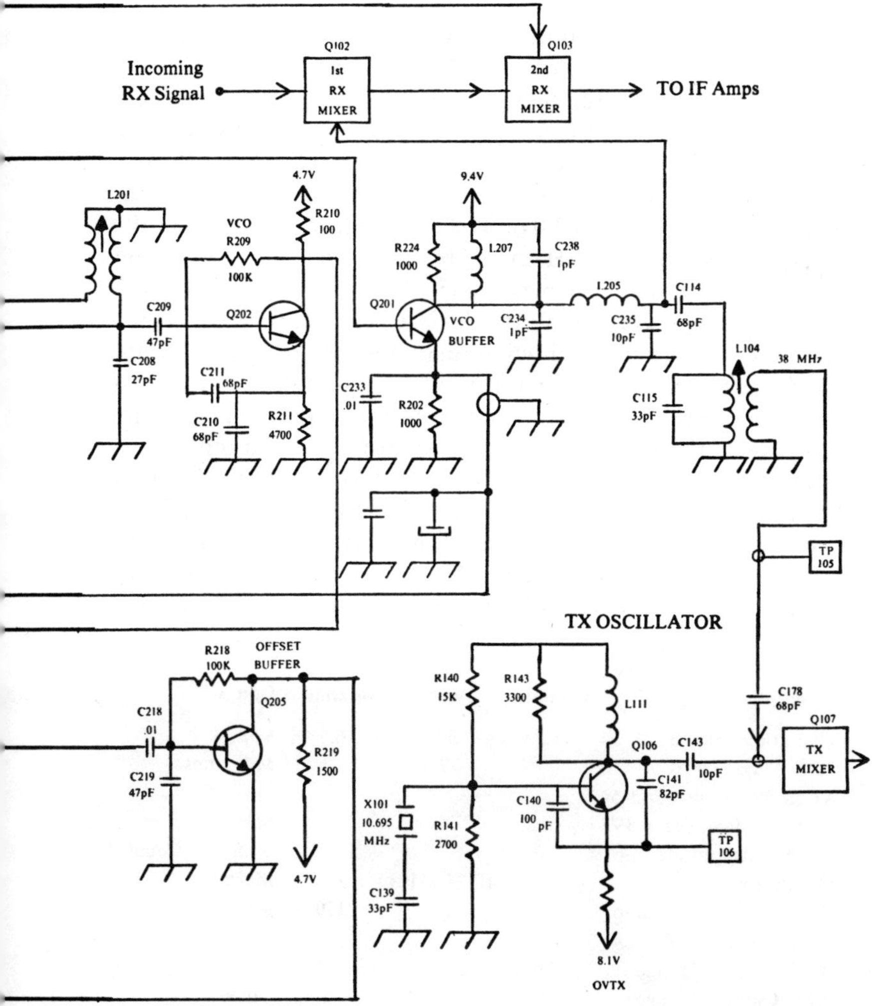
Incoming
RX Signal
Q102
1st
RX
MIXER
Q103
2nd
RX
MIXER
TO IF Amps
L201
4.7V
R210
100
VCO
R209
100K
C209
47pF
Q202
C208
27pF
C211
68pF
C210
68pF
R211
4700
9.4V
R224
1000
L207
C238
1pF
Q201
VCO
BUFFER
C234
1pF
L205
C235
10pF
C114
68pF
L104
38 MHz
C115
33pF
C233
.01
R202
1000
TP
105
TX OSCILLATOR
R218
100K
OFFSET
BUFFER
Q205
C218
.01
C219
47pF
R219
1500
4.7V
R140
15K
R143
3300
L111
Q106
C143
10pF
C178
68pF
Q107
TX
MIXER
C141
82pF
C140
100 pF
X101
10.695
MHz
R141
2700
TP
106
C139
33pF
8.1V
OVTX

line open for binary 0. If a program line is left open, the on-chip pulldown resistor automatically returns the program pin to binary 0.

Using channel 1 as the example, here is the analysis of this PLL circuit operation. Refer to Fig. 4-4, 4-5, and Table 4-3 during this discussion. As mentioned previously in this chapter, the VCO operates in the 37 to 38 MHz range.

Assume that the channel selector is set for channel 1 and then power is applied to the CB set. Table 4-3 shows the VCO frequency to be 37.660 MHz for channel 1. When the circuit is first energized, the VCO starts operating at some random frequency in the 37 to 38 MHz vicinity. For this example the VCO initially starts operating at 37 MHz. This will make the input to the programmable divider 37.00 - 36.38, or 0.62 MHz. The programmable divider divides this 0.62 MHz signal by 128 to yield 4.84 kHz at the output of the programmable divider. This 4.84-kHz signal at the output of the programmable divider is fed to the input of the automatic frequency control circuit (afc) that is within IC202. The afc compares the 4.84-kHz signal from the programmable divider to the 10-kHz reference signal from the reference divider. The afc circuit detects the large frequency difference of its two inputs and produces an error signal, which is applied to pin 1 of the IC. This error signal is in the form of positive-going pulses. These positive-going pulses are filtered by an RC filter network to form a relatively pure DC control voltage. The filter network consists of R206, R207, C205, and C206. The control voltage is applied to the

Table 4-3. Fanon Fanfare 190 DF PLL Specifications.

Description	Minimum	Typical	Maximum	Unit
Operating voltage	4.5	5.0	6.0	Volts
IDD Max(FQ=10.24MHz VCOin=2.5MHz, VDD=5v		2		Milliamps
VCO in Max (VDD=5V)		4		MHz
VCO in Amplitude (p-p)	2.0			Volts (AC Coupled)
Quartz Frequency		10.24 MHz		Megahertz
Current output to ramp capacitor (pin 18)	70	90	110	Microamps
APC Output Voltage	2.0		5.3	Volts
AFC Output Voltage	0		6	Volts
Reference Crystal				
Load Capacity		32		Picafarad
Impedance			40	Ohms
Lock Detector Load	20,000			Ohms

VCO control line (at one end of R208). This control voltage then moves the VCO frequency upward. As the VCO frequency increases, the input to the programmable divider approaches 1.28 MHz and the output of the programmable divider approaches 10 kHz. When the VCO frequency reaches 37,660 MHz, the input to the programmable divider wili be 37.660 MHz - 36.380 MHz, or 1.28 MHz. The output of the programmable divider will be 1.28 MHz ÷ 128, or 10 kHz.

At this point, the lock detector indicates a frequency acquisition and therefore produces a HIGH level voltage at pin 8 of the IC. This HIGH level voltage is also applied to an APC switch (see Fig. 4-1), which enables the phase detector output to appear at pin 7 of the IC. This phase detector output serves to fine tune the VCO to keep it on track.

The 37.660-MHz VCO signal is buffered by Q201 and then fed to the receive first mixer where it is heterodyned with the incoming channel 1 signal to produce a difference frequency of 37.660 MHz - 26.965 MHz, or 10.695 MHz. This 10.695-MHz signal is the first intermediate frequency. This first i-f is then fed to the second receive mixer where it is heterodyned with a 10.240-MHz signal from the reference oscillator buffer. The difference frequency, 10.695 MHz - 10.240 MHz = 455 kHz, appears at the output of the second mixer. This 455-kHz second i-f signal is then amplified and detected by the following stages.

In the transmit mode, the 37.660-MHz VCO signal is fed from the secondary of L104 to the input (base) of the transmit mixer transistor, Q107. Also, a 10.695-MHz signal from the transmit oscillator is fed to the base of Q107. These two signals are heterodyned to produce a difference frequency of 37.660 MHz - 10.695 MHz, or 26.965 MHz. This is the correct channel 1 frequency. Notice that the mixing process for the transmit mode is just the opposite of the RX mode.

Figure 4-5 shows that the forward bias for the VCO buffer transistor, Q201 is supplied from pin 8 of IC202 through diode D202 and resistors R204 and R203. Pin 8 is the lock detector output. Under normal locked conditions, the voltage at pin 8 is at a HIGH level; thus, Q201 is forward biased. When an unlocked condition exists, the voltage at pin 8 falls to a LOW level. This kills the forward bias to the VCO buffer stage, so the output of the synthesizer is killed. This disables both the receiver and the transmitter. It is most important for the transmitter to be killed during unlocked conditions to prevent off-frequency signals from being radiated.

One slightly unusual feature of this radio is that the channel selector switch and the LED channel displays are located inside the microphone itself. Figure 4-6 shows a diagram of the circuitry contained in the microphone head. The microphone cable contains 12 wires which connect to the radio through a 12-pin plug. The group of diodes (a through g) at the top in Fig. 4-6 represents the units LED display. The group of diodes (a through g) at the bottom represents the tens LED display. Figure 4-7 shows the connections of the various terminal pins of the LED displays. Each segment is supplied approximately 1.9 volts and draws about 10 mA.

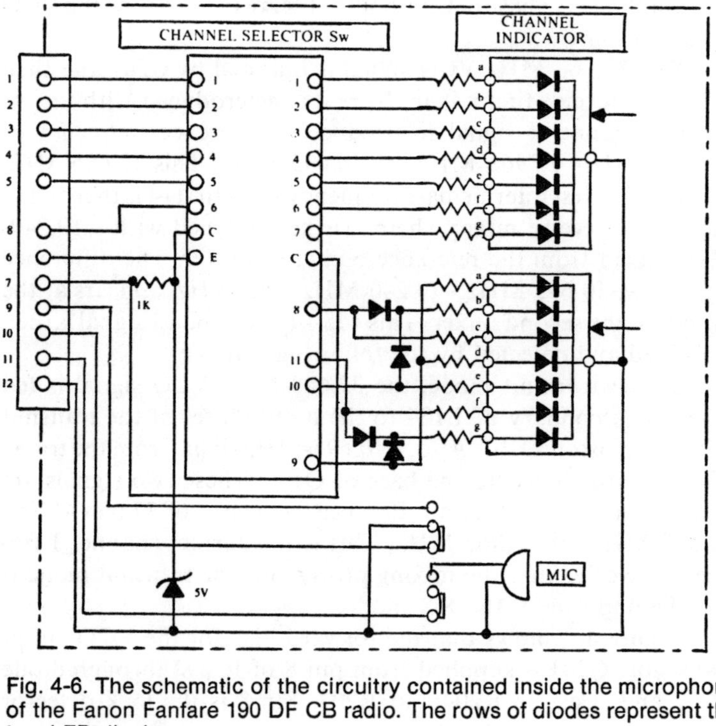

Fig. 4-6. The schematic of the circuitry contained inside the microphone of the Fanon Fanfare 190 DF CB radio. The rows of diodes represent the two LED displays.

Modular Construction

This PLL circuit is built on a separate circuit board module which connects to the main circuit board through two plugs. The module is housed in a metal box. To perform alignment or repair on the PLL unit, the module must be removed from the metal box and then plugged back into the main board for live tests. When doing this, be sure to insulate the PLL module with cardboard or some other insulating material to prevent shorting the module during live testing. to remove the PLL module for servicing, perform the following steps:

☐ Remove the shield from the side chassis.

☐ Remove the three PLL module retaining screws from the chassis.

☐ Disconnect the two plugs-making certain that the lock tabs release—and remove the PLL module from the chassis.

☐ Remove the two retaining screws and then the cover from the PLL module.

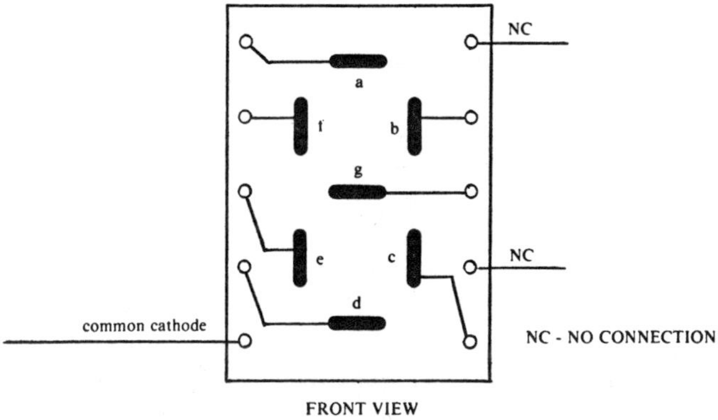

Fig. 4-7. Front view of one of the LED displays used in the Fanon Fanfare 190 DF for channel indication.

Alignment

☐ Connect a frequency counter to TP1 and check for 10.240 MHz. There is no adjustment to correct this.

☐ With the channel selector on channel 19, connect an rf voltmeter to TP4 and adjust L202 for maximum rf voltage.

☐ Connect a frequency counter to TP4 and check for 36.380 MHz. The frequency should be within + 100 Hz. If necessary, touch up L202.

☐ Connect a DC voltmeter to TP3. On channel 19, the voltage at TP3 should be 3.0 volts. If not, adjust L201 for 3.0 volts.

☐ Using a rf voltmeter, adjust L104 for maximum rf voltage at TP105.

☐ Connect a frequency counter to TP105 and check for the proper VCO frequency on each channel. See Table 4-3 for the VCO frequency for each channel.

☐ With the channel selector on channel 20, key the transmitter and adjust L111 for maximum rf voltage at TP106.

☐ With the channel selector on channel 20, connect a frequency counter to TP106. Key the transmitter and check for 10.695 MHz. If necessary, touch up L111.

☐ Key the transmitter and make certain that the carrier frequencies are correct to within + 100 Hz for all channels.

Troubleshooting

If the set neither transmits nor receives on any channel, the first troubleshooting step would be to check the supply voltage to the PLL circuit. The convenient place to check the voltage is at pin 2 of IC202. If the supply voltage is missing, troubleshoot the power supply. The voltage for the PLL section is supplied from a special 5-volt regulator IC. This regulator IC is IC201. It is located on the PLL board itself. Figure 4-8 shows the diagram of the regulator circuit which supplies the PLL circuit. The service manual states that the IC is either a muPC14305 or an MC78L05. If the supply voltage is present at pin 2 of IC202, check for rf voltage at TP105. If no rf voltage is present at TP105, check the DC voltage level at pin 8 of IC202. This is the lock detector output. If the DC voltage at pin 8 is at a LOW level, the loop is not locked. Thus, the forward bias to the VCO buffer (Q201) is killed. An unlocked condition could be caused by several different things. Some possibilities are missing VCO signal, missing offset oscillator signal (36.380 MHz), offset oscillator operating far off frequency, offset mixer or offset buffer malfunction, VCO misadjusted or possibly a defective IC202.

On the other hand, if the DC voltage at pin 8 is at a HIGH level (near 5 volts), the PLL loop is locked. In this case, use your rf voltmeter to check for rf voltage at the base of Q201 and then the collector of Q201 on through the output coupling network.

Wrong Transmit and Receive Frequencies on High Channels with Low Channels OK. This is usually caused by improper adjustment of the VCO oscillator. This would probably result in a higher than normal control voltage at TP3. For VCO adjustment use the fourth step of the alignment instructions.

Wrong Transmit and Receive Frequencies on Low Channels with High Channels OK. This is usually caused by improper adjustment of the VCO oscillator. This would probably result in a lower than normal control voltage at TP3. For VCO adjustment see the fourth step of the "Alignment" instructions.

Channel NO.	VCO FREQ. (MHz)	Divider INPUT (MHz)	Divisor PIN	Ic Programming Code							
				9	10	11	12	13	14	15	16
				128	64	32	16	8	4	2	1
1	37.660	1.28	128	1	0	0	0	0	0	0	0
2	37.670	1.29	129	1	0	0	0	0	0	0	1
3	37.680	1.30	130	1	0	0	0	0	0	1	0
4	37.700	1.32	132	1	0	0	0	0	1	0	0
5	37.710	1.33	133	1	0	0	0	0	1	0	1
6	37.720	1.34	134	1	0	0	0	0	1	1	0
7	37.730	1.35	135	1	0	0	0	0	1	1	1
8	37.750	1.37	137	1	0	0	0	1	0	0	1
9	37.760	1.38	138	1	0	0	0	1	0	1	0
10	37.770	1.39	139	1	0	0	0	1	0	1	1
11	37.780	1.40	140	1	0	0	0	1	1	0	0
12	37.800	1.42	142	1	0	0	0	1	1	1	0
13	37.810	1.43	143	1	0	0	0	1	1	1	1
14	37.820	1.44	144	1	0	0	1	0	0	0	0
15	37.830	1.45	145	1	0	0	1	0	0	0	1
16	37.850	1.47	147	1	0	0	1	0	0	1	1
17	37.860	1.48	148	1	0	0	1	0	1	0	0
18	37.870	1.49	149	1	0	0	1	0	1	0	1
19	37.880	1.50	150	1	0	0	1	0	1	1	0
20	37.900	1.52	152	1	0	0	1	1	0	0	0
21	37.910	1.53	153	1	0	0	1	1	0	0	1
22	37.920	1.54	154	1	0	0	1	1	0	1	0
23	37.950	1.57	157	1	0	0	1	1	1	0	1
24	37.930	1.55	155	1	0	0	1	1	0	1	1
25	37.940	1.56	156	1	0	0	1	1	1	0	0
26	37.960	1.58	158	1	0	0	1	1	1	1	0
27	37.970	1.59	159	1	0	0	1	1	1	1	1
28	37.980	1.60	160	1	0	1	0	0	0	0	0
29	37.990	1.61	161	1	0	1	0	0	0	0	1
30	38.000	1.62	162	1	0	1	0	0	0	1	0
31	38.010	1.63	163	1	0	1	0	0	0	1	1
32	38.020	1.64	164	1	0	1	0	0	1	0	0
33	38.030	1.65	165	1	0	1	0	0	1	0	1
34	38.040	1.66	166	1	0	1	0	0	1	1	0
35	38.050	1.67	167	1	0	1	0	0	1	1	1
36	38.060	1.68	168	1	0	1	0	1	0	0	0
37	38.070	1.69	169	1	0	1	0	1	0	0	1
38	38.080	1.70	170	1	0	1	0	1	0	1	0
39	38.090	1.80	180	1	0	1	0	1	0	1	1
40	38.100	1.81	181	1	0	1	0	1	1	0	0

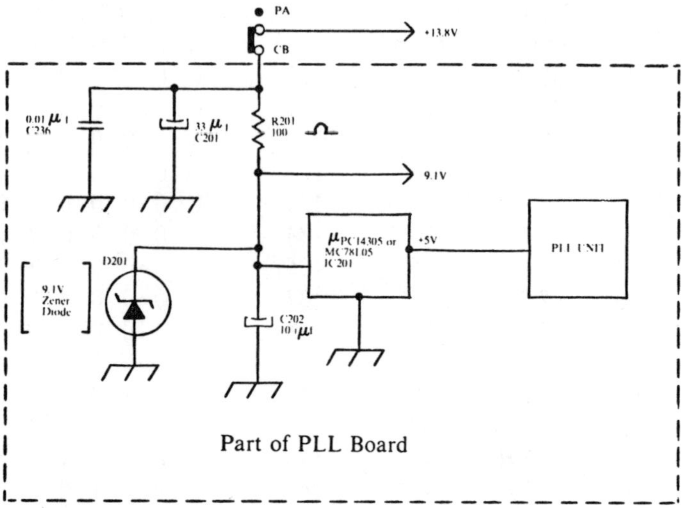

Fig. 4-8. Partial schematic which shows the voltage regulator circuit which supplies the PLL circuit.

Weak or No Receive and Off-Frequency Transmit on All Channels. This is usually the result of incorrect VCO frequency, although the VCO itself is probably not at fault. If the offset oscillator is operating off frequency, this would in turn cause the VCO frequency to be wrong. This in turn affects both the receive and transmit signals.

Off-Frequency Transmit on All Channels with Receive OK. This could be caused by the transmit oscillator operating off frequency. Use your frequency counter and check at TP106 to make certain that the 10.695-MHz signal is correct at that point.

Some Channels Inoperative. Check the channel switch. Check the program lines to IC202 for the proper binary levels. Don't overlook the possibility of a defective IC202.

Some Channels Intermittent. This is usually caused by dirty or loose contacts on the channel selector switch. Clean the selector switch with a good grade of tuner cleaner.

Chapter 5

The PLL02A Digital Synthesizer IC

The PLL02A is a CMOS LSI integrated circuit. This IC is used in many different brands of CB radios. The IC contains a fixed frequency divider, programmable divider, phase detector, and charge pump. The IC is housed in a 16-pin dual-in-line package.

SPECIAL FEATURES

- ☐ CMOS design for low-power dissipation
- ☐ Pure binary programming code
- ☐ On-chip pull-down resistors at programming inputs
- ☐ Wide operating temperature range
- ☐ Lock-detector output for stopping TX output when loop is unlocked
- ☐ 16 pin dual-in-line package

INTERNAL STRUCTURE

A block diagram of the internal structure of the PLL02A is shown in Fig. 5-1. Figure 5-2 shows the pin numbering scheme and the pin functions of the IC.

Pin number 1 is the supply voltage input for the IC. The IC supply voltage is supplied from a special voltage regulator to hold the supply voltage constant to ensure more stable operation of the PLL circuit. Pin 2 is the input to the programmable divider. Pin 3 is the input to the fixed reference divider. The input to pin 3 would be 10.240 MHz. This signal is supplied from an external crystal-

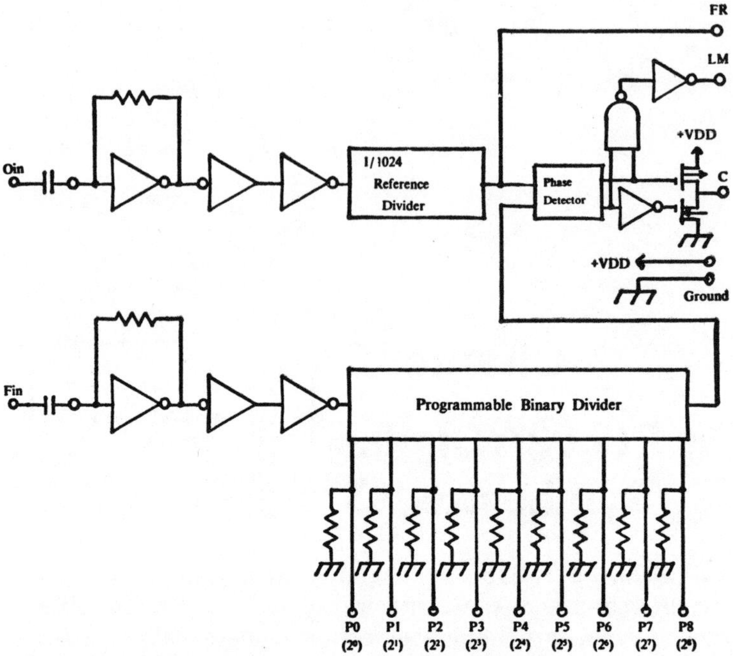

Fig. 5-1. Internal view of the PLL02A IC.

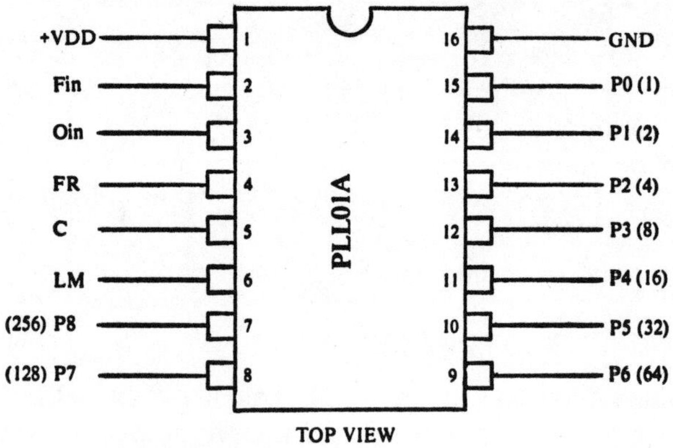

+VDD—+5V
Fin—Programmable divider input
Oin—Reference divider input
FR—10 kHz output
C—Charge pump output
LM—Lock monitoring output
P8 P0—Programming lines of programmable divider
GND—Ground

Fig. 5-2. The pin numbering scheme and pin functions for the PLL02A.

controlled reference oscillator. The fixed divider divides this oscillator signal by 1024 to yield a 10-kHz reference frequency for the phase detector. Pin 4 is connected to the output of the reference divider. Pin 5 is the output of the charge pump. This output has to be filtered before being applied to the VCO control line. Pin 6 is the lock-detector output. Under normal locked conditions, this pin is at a HIGH level. When an unlocked condition occurs, the voltage on the pin drops to a lower level. This can be used to stop the transmitter from radiating during conditions of unlock. Pins 7 through 15 are the programming lines of the programmable divider. The programming is done in the pure binary form. If a program line is connected to VDD (+5 volts), it is at binary 1 level. If the line is left open, the pulldown resistor pulls the program line to ground potential, thereby placing the line at the binary 0 level. If all of the program input lines are at binary 1, the divider will be programmed to divide its input frequency by 511. This is the maximum divisor (N) that can be programmed into the divider. To determine the divisor (N) which is programmed into the programmable divider, simply add the weights of the program pins where the binary 1 level appears. The weights of the various program pins are shown in Fig. 5-2. Notice that pin number 15 is the least significant bit while pin number 7 is the most significant bit.

Table 5-1 shows how the divisor (N) of 341 is programmed into the programmable divider. The weight of each pin is listed below that pin number. The sum of the weights where binary 1 appears is equal to the decimal equivalent of the binary-coded number. Pin 16 is the common or ground terminal of the IC. This is usually labeled as VSS on CMOS devices.

Table 5-2 lists the absolute maximum ratings of the PLL02A IC. Tables 5-3 through 5-6 list the electrical characteristics of the PLL02A.

Table 5-1. PLL02A PLL Weights and Programming Codes.

	Pure Binary Code									Decimal Equivalent
Pin Number	7	8	9	10	11	12	13	14	15	
Pin Weight	256	128	64	32	16	8	4	2	1	
Binary Level	1	0	1	0	1	0	1	0	1	341

Table 5-2. Absolute Maximum Ratings of the PLL02A.

ITEM	SYMBOL	RATING	UNIT
Power supply voltage	VDD	-0.3---7	V
Input voltage	Vin	-0.3--VDD+0.3	V
Operating temperature	TOP	-30--+70	°C
Storage temperature	TST	-40--+85	°C

Table 5-3. PLL02A Supply Voltage and Current Rating.

Item	Symbol	Test Conditions	Min	Typ	Max	Units
Operating voltage range	VDD	Ta=-30--+70° C	4.5	5.0	5.5	V
Current consumption	IDD	VDD = 5.5 V			15	MA

Ta = Ambient temperature

Table 5-4. Programmable Counter Specifications.

Item	Symbol	Test Conditions	Min	Typ	Max	Unit
Input frequency	Ff. in	Sine wave input VDD = 4.5 V	4.5			MHz
Input voltage, Vp-p	Vf. in	Sine wave input VDD = 4.5 V	0.5			V
Pull-down resistance P0-P8	RP	+25 C	10	20	50	K

Table 5-5. Standard Frequency Divider Specifications.

Item	Symbol	Test Conditions	Min	Typ	Max	Unit
Input frequency	Fq. in	Sine wave input VDD = 4.5 V		10.24		MHz
Input voltage, Vp-p	Vq. in	Sine wave input VDD = 4.5 V	1.0			V

Table 5-6. Charge Pump Output and Lock Monitoring Output Specifications.

Item	Symbol	Test Conditions	Min	Typ	Max	Unit
Charge-pump output voltage	Vc. h	VDD = 5 V IDD. h=0.5MA	4.0			V
	Vc. 1	VDD = 5 V IDD. 1=0.5MA			1.0	V
Lock-monitoring output voltage	V1. h	VDD = 5 V IDD. h=0.7MA	4.0	4.5		V

PRACTICAL APPLICATION

Figure 5-3 shows a block diagram of a typical application of the PLL02A digital synthesizer IC. This particular application of the PLL02A is used in several 23-channel transceivers, including the Midland 13-882C. Pins 7 through 15 of the PLL02A are the programming pins. On the block diagram, you will notice that only pins 11, 12, 13, 14, and 15 are connected to the channel switch. The other programming pins — 7, 8, 9, and 10 — are con-

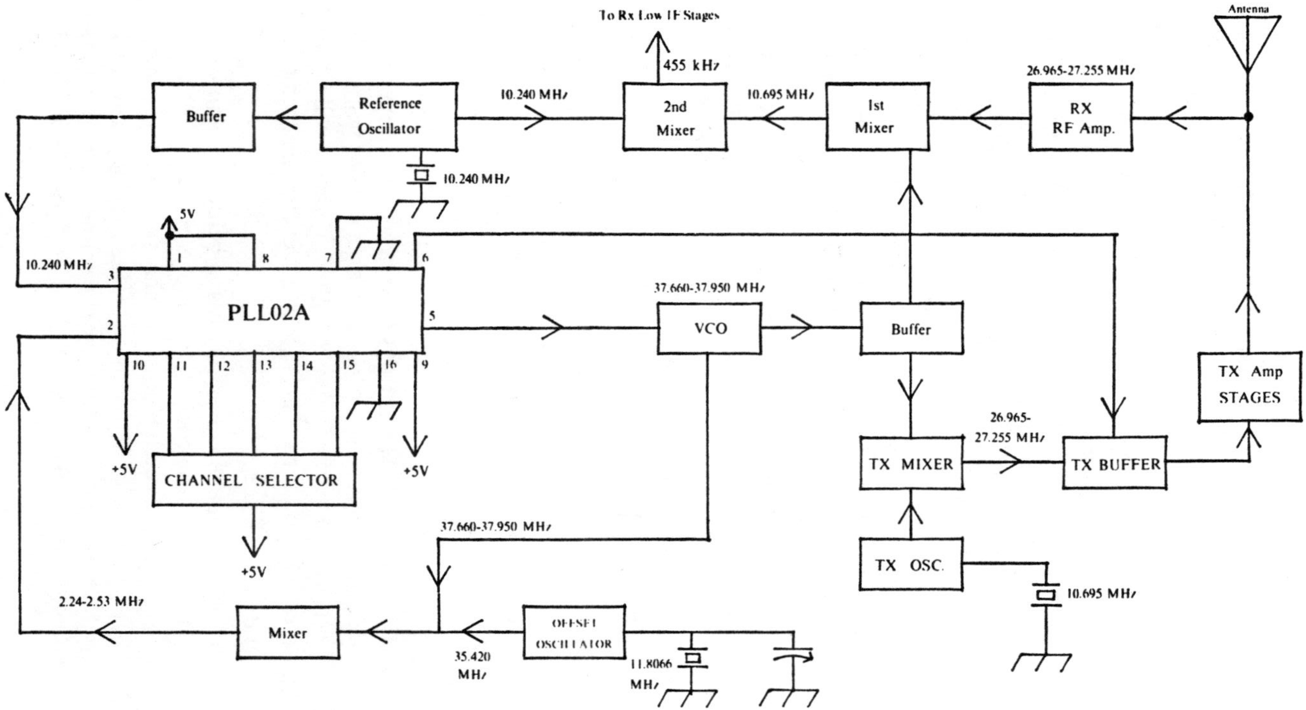

Fig. 5-3. Block diagram of typical application of the PLL02A IC. This circuit was widely used in 23-channel PLL circuits.

nected to a fixed binary level. Pin 7 (which has a weight of 256) is connected directly to ground or binary 0. This means that the divisor N of the programmable divider will always be less than 256. Pins 8, 9, and 10 are connected to +5 volts or binary 1. The weights of pins 8, 9, and 10 are 128, 64, and 32, respectively. The sum of these weights is 128 + 64 + 32, or 224. Thus, the minimum divisor N is set at 224. So the divisor N range is 224 through 255. If all the program lines connected to the channel selector switch were placed at a HIGH level (binary 1) by the selector, the divisor N will be 255. If all of the program lines connected to the channel selector were left open, the divisor will be 224.

Table 5-7 lists the N divisor, programmable divider input frequency, VCO frequency, and programming code for each of the 23 channels. Using channel 1 as the example, here is how this synthesizer works. The VCO operates in the 37-MHz range. When the channel selector is set for channel 1, the programmable divider is programmed to divide by 224 (see Table 5-7). The reference frequency for the frequency/phase detector is 10 kHz, so when the output of the programmable divider is 10 kHz, the loop will lock up. In order for the output of the programmable divider to be 10 kHz, the input to the divider must be 10 kHz × 224, or 2.24 MHz. Notice on the block diagram that the input to the divider is taken from the output of a mixer which mixes the VCO frequency with the offset oscillator frequency. The offset oscillator operates at the third overtone of an 11.806666-MHz crystal to yield a 35.420-MHz signal. Thus, the output of the mixer is fVCO (VCO frequency) - 35.420 MHz.

The proper VCO frequency for channel 1 is 35.420 MHz + 2.240 MHz, or 37.660 MHz. If the VCO frequency is not exactly 37.660 MHz, the input to the programmable divider will not be 2.24 MHz. In this case, the frequency/phase detector will issue a correction voltage to correct the VCO frequency. The correction voltage appears at pin 5 of the PLL02A.

Actually, the control voltage is filtered before it is applied to the VCO control line. The correction voltage moves the VCO frequency in the proper direction until it reaches 37.660 MHz. When the VCO frequency reaches 37.660 MHz, the loop will lock. Notice on the block diagram that pin 6 is connected to the transmit buffer stage. When the loop is locked, the voltage on pin 6 is normally at a HIGH level. This voltage is used to supply forward bias to the transmit buffer stage. When an unlocked condition oc0 curs, the voltage at pin 6 drops to a lower level. This then kills the

Table 5-7. Specifications for the PLL Circuit of Fig. 5-3.

Channel No.	"N" Divisor	Prog. Div. Input Freq.	VCO Freq.	IC PIN NO 11	12	13	14	15
1	224	2.24MHz	37.660 MHz	0	0	0	0	0
2	225	2.25 "	37.670 "	0	0	0	0	1
3	226	2.26 "	37.680 "	0	0	0	1	0
4	228	2.28 "	37.700 "	0	0	1	0	0
5	229	2.29 "	37.710 "	0	0	1	0	1
6	230	2.30 "	37.720 "	0	0	1	1	0
7	231	2.31 "	37.730 "	0	0	1	1	1
8	233	2.33 "	37.750 "	0	1	0	0	1
9	234	2.34 "	37.760 "	0	1	0	1	0
10	235	2.35 "	37.770 "	0	1	0	1	1
11	236	2.36 "	37.780 "	0	1	1	0	0
12	238	2.38 "	37.800 "	0	1	1	1	0
13	239	2.39 "	37.810 "	0	1	1	1	1
14	240	2.40 "	37.820 "	1	0	0	0	0
15	241	2.41 "	37.830 "	1	0	0	0	1
16	243	2.43 "	37.850 "	1	0	0	1	1
17	244	2.44 "	37.860 "	1	0	1	0	0
18	245	2.45 "	37.870 "	1	0	1	0	1
19	246	2.46 "	37.880 "	1	0	1	1	0
20	248	2.48 "	37.900 "	1	1	0	0	0
21	249	2.49 "	37.910 "	1	1	0	0	1
22	250	2.50 "	37.920 "	1	1	0	1	0
23	253	2.53 "	37.950 "	1	1	1	0	1

forward bias to the transmit buffer stage, thus preventing the transmitter from radiating improper frequencies.

On channel 1, the 37.660-MHz signal is fed to the VCO buffer and then to the first mixer. In the first mixer, the 37.660-MHz signal is heterodyned with the incoming 26.965-MHz, channel 1 signal to yield a difference frequency of 37.660 MHz - 26.965 MHz, or 10.695 MHz at the output of the first mixer. This 10.695-MHz signal is the first intermediate frequency. The signal is then fed to second mixer, where it is heterodyned with a 10.240-MHz signal from the reference oscillator. The output of the second mixer is 10.695 Mhz - 10.240 MHz, or 455 kHz. This is the *second* or *low* intermediate frequency. This 455-kHz i-f is then amplified and detected by the following receiver stages.

In the transmit mode, the 37.660-MHz VCO signal is fed to the transmit mixer, where it is heterodyned with a 10.695-MHz signal from the transmit oscillator. The difference of these two frequencies is 37.660 MHz - 10.695 MHz, or 26.965 MHz. This is the proper transmit frequency for channel 1. This signal is then amplified by the transmit buffer and following transmitter stages before being applied to the antenna.

MIDLAND 77-830 CB PLL SYNTHESIZER

The Midland model 77-830 CB radio is shown in Fig. 5-4. This is a 40-channel CB set, one of many which used the PLL02A digital PLL IC. The block diagram of the PLL section of the transceiver is shown in Fig. 5-5. The full schematic of the PLL synthesizer is shown in Fig. 5-6.

Reference Oscillator Stage

The reference oscillator stage is composed of transistor Q1 and its associated components. Also, transistor Q2 serves as a buffer between the oscillator and the input to the reference divider, so Q2 can be considered to be part of the reference oscillator stage. Crystal X1 resonates the oscillator at 10.240 MHz. Trimmer capacitor CT1, which is connected between one side of crystal X1 and ground, serves as a fine tuning control to adjust the oscillator to exactly 10.240 MHz. A sample of the 10.240-MHz signal is taken from the emitter of Q1 and fed to transistor Q2. The buffer stage is connected as an emitter follower. The 10.240-MHz signal is taken from the emitter of Q2 and fed through capacitor C61 to the input of the reference frequency divider (divide by 1024) at pin 3 of IC1, the PLL02A. A portion of the 10.240-MHz signal is also fed to the receiver second mixer. This signal is taken from the junction of CT1 and X1. At the second mixer, the 10.240-MHz signal is heterodyned with the 10.695-MHz high intermediate frequency to yield a low intermediate frequency of 455 kHz.

At the collector of transistor Q1 (reference oscillator), the primary of transformer T1 serves as a load in the collector circuit. The secondary of this transformer is tuned to 20.480 MHz, the second harmonic of the 10.240-MHz oscillator signal. This 20.480-Mhz signal is fed to pin 4 of the VCO/mixer, IC-2. The function of this IC will be discussed next.

VCO and Mixer

The VCO and mixer is a C3001A. Shown in the schematic as IC2, this is a nine-pin IC. Figure 5-7 shows a drawing of the C3001A and the pin numbering scheme. The IC is essentially an rf amplifier and dual-balanced mixer. The rf amplifier serves as the oscillator when external components are used to resonate the

Fig. 5-4. The Midland 77-830 CB radio (courtesy of Midland International Corp.).

oscillator and an external capacitor used to provide feedback for oscillation.

Look at the PLL schematic of Fig. 5-6. At IC2, the VCO and mixer, a 220-pF capacitor (C7) connects between pins 1 and 2 of the IC. The capacitor provides feedback for the voltage-controlled oscillator. the resonant frequency of the VCO is determined by varactor diode D1 and coil L1 with its parallel capacitor. Also, capacitors C4, C6, C7, and C8 affect the resonant frequency to some extent. Notice that the cathode of the varactor diode is connected to the 5.4-volt supply through a 47K ohm resistor, R4. The anode of the varactor diode connects to the VCO control line through resistors R3 and R1. The voltage on the control line is lower than the voltage on the cathode of the varactor. This reverse biases the varactor. In most of the VCOs used in PLL applications, the anode of the varactor connects to ground and the cathode connects to the VCO control line. In this application, however, the anode connects to the control line and the cathode is placed at rf ground by the 0.01-muF capacitor, C5. The VCO operates in the range of 17.180 MHz for channel 1 to 17.620 MHz

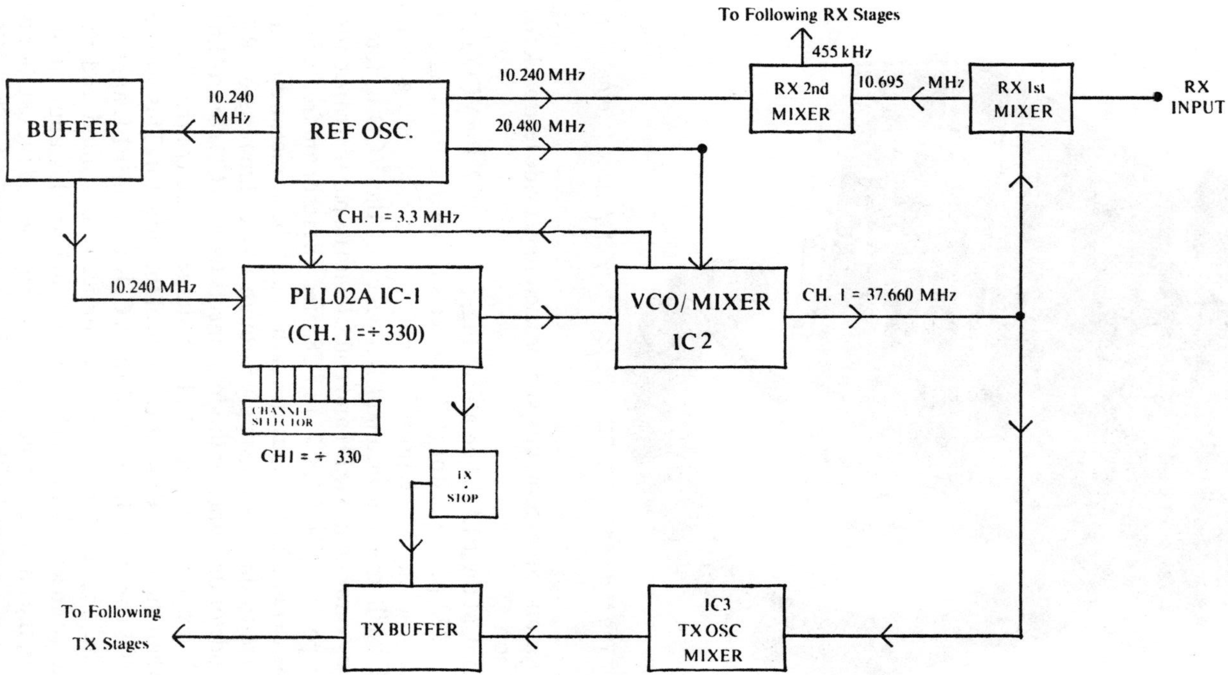

Fig. 5-5. Block diagram of the PLL synthesizer used in the Midland 77-830. The frequencies listed at various points on the schematic are for channel 1 operation.

for channel 40.

Pin 4 of IC2 is a mixer input. A 20.480-MHz signal, which is derived from doubling the reference oscillator frequency, is applied to pin 4. Inside the IC, this 20.480-MHz signal is mixed with the 17-MHz signal from the VCO to yield a sum and a difference frequency. The sum frequency appears at pin 6 of the IC. The sum frequency ranges from 37.660 MHz for channel 1 to 38.100 MHz for channel 40. The difference frequency appears at pin 9 of the IC. The difference frequency ranges from 3.30 MHz for channel 1 to 2.86 MHz for channel 40.

Transmit Oscillator/Mixer

The transmit oscillator and mixer (IC3) uses the same type of IC as the VCO and mixer. In Fig. 5-6, the oscillator frequency is determined by crystal X2. This crystal sets the oscillator frequency at 10.695 MHz. Capacitor C26, at 560 pF, provides feedback for the oscillator. The sum frequency signal from the VCO and mixer is applied to IC3 at pin 4. This frequency at pin 4 ranges from 37.660 MHz for channel 1 to 38.100 MHz for channel 40. Inside IC3 this 37-Mhz signal at pin 4 is mixed with the 10.695-MHz oscillator signal. The difference of the two frequencies appears at pin 9 of IC3. The frequency at pin 9 ranges from 26.965 MHz for channel 1 to 27.405 MHz for channel 40. This signal is then fed to the following transmitter stages for amplification.

Transmit Stop Circuit

The transmit stop circuit is composed of transistor Q22 and associated components. Actually, the signal that operates the transmit stop circuit comes from the lock detector that is part of IC1, the PLL02A. This circuit prevents the radiation of improper frequencies from the transmitter in the event the loop cannot achieve lockup. The circuit works as follows: Under normal locked conditions the voltage at pin 6 of IC1 (the lock detector output of the PLL02A) is at a HIGH level (typically 4.5 volts). Transistor Q22 is simply a switch that when turned on, supplies forward bias to transmit buffer transistor Q3 through resistor R17. Normally, voltage at the base of Q22 is approximately 5 volts and the voltage at pin 6 of IC1 is normally approximately 4.5 volts. This means that diode D22 will be reverse biased under normal locked conditions.

However, if an unlocked condition occurs, the voltage at pin 6 of IC1 will go to a lower value. Then diode D22 will be forward

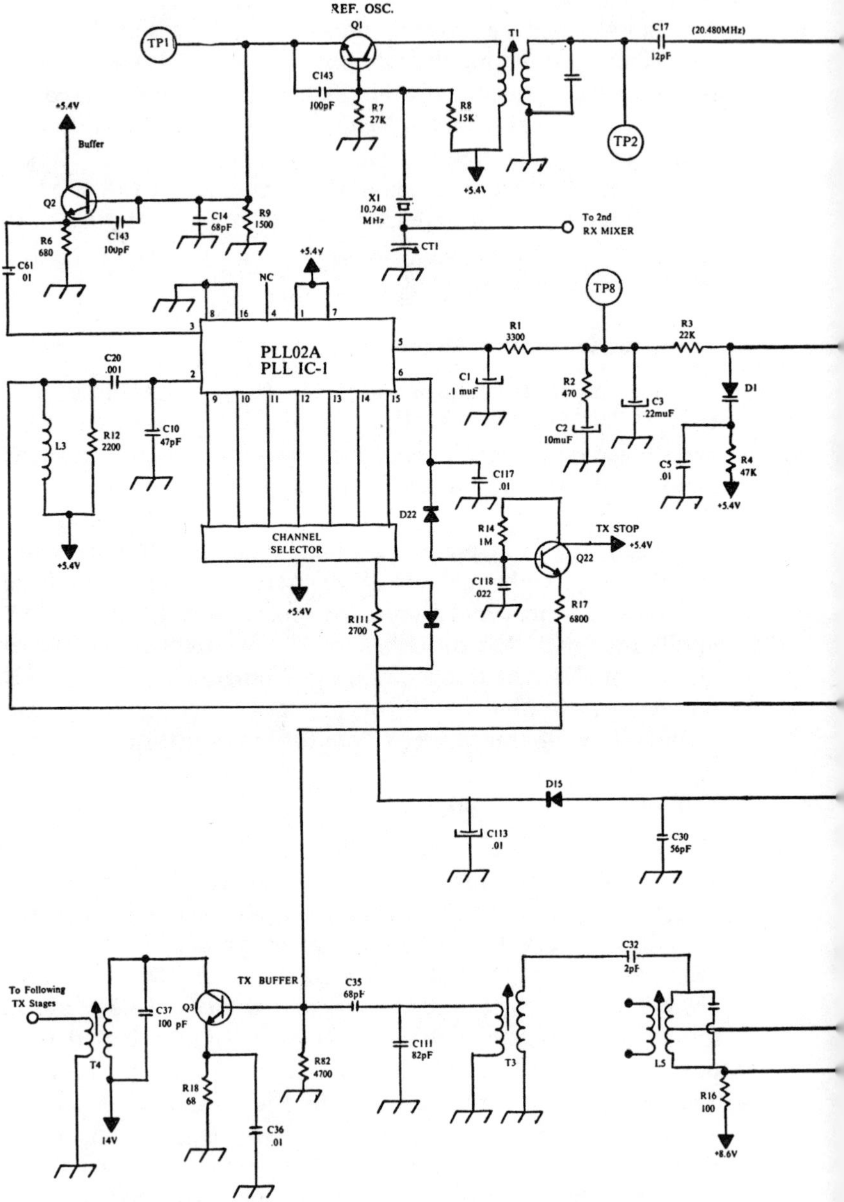

Fig. 5-6. Midland 77-830 PLL synthesizer.

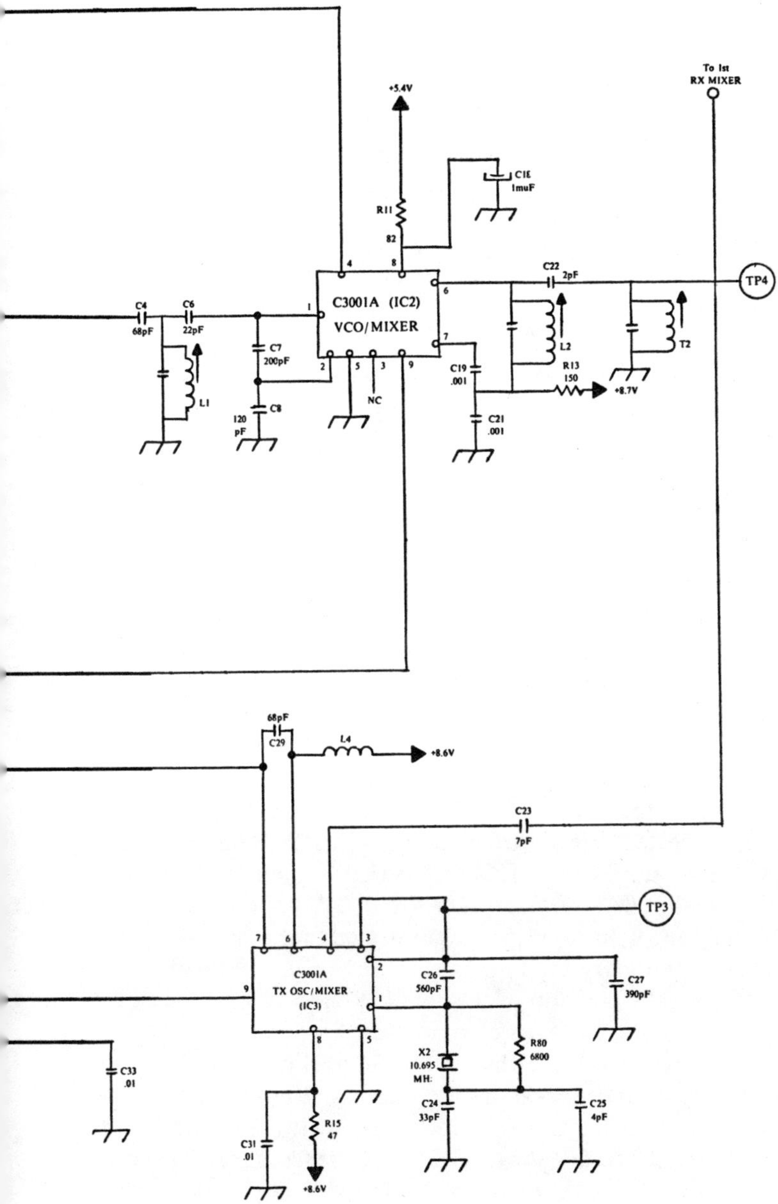

To Ist
RX MIXER
+5.4V
C18
1muF
R11
82
4
8
C22
2pF
TP4
6
C4
68pF
C6
22pF
1
C3001A (IC2)
VCO/MIXER
7
L2
T2
C7
200pF
2
5
3
9
C19
.001
R13
150
+8.7V
L1
C8
120
pF
NC
C21
.001
68pF
C29
L4
+8.6V
C23
7pF
TP3
7
6
4
3
C3001A
TX OSC/MIXER
(IC3)
2
C26
560pF
C27
390pF
9
1
8
5
R80
6800
X2
10.695
MH:
C33
.01
C24
33pF
C25
4pF
R15
47
C31
.01
+8.6V

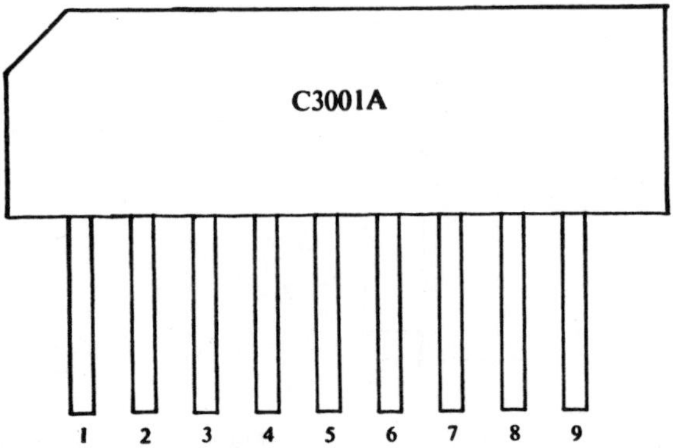

Fig. 5-7. The C3001A IC pin numbering scheme.

biased, and the base of Q22 will be pulled down to a LOW level, thus killing the forward bias to Q22. This action causes Q22 to cut off. This cutoff in turn removes the forward bias from the transmit buffer, thus stopping the transmitter.

Digital Synthesizer

The PLL02A, IC1 in Fig. 5-6, performs several functions. First, let's take a look at the programming used in this particular application. The programming pins are 7, 8, 9, 10, 11, 12, 13, 14, and 15. Table 5-1 shows the weight of each programming pin. In Fig. 5-6, pin 7 is connected to the supply voltage. This keeps it at the binary 1 level at all times, regardless of the channel selector position. Thus, the weight of pin 7 (256) will always be used. Also, pin 8 is connected directly to ground, so this pin remains at the binary 0 level at all times. This means that its weight of 128 will never be used in determining the N divisor. All of the other programming pins (9 through 15) are connected to the channel selector switch. the binary level applied to any of pins 9 through 15 depends upon the position of the channel selector. There is a different combination for each channel (see Table 5-8).

Notice from Table 5-8 that as the channel numbers increase from 1 to 40, the divisor N actually decreases. Why is this done? In most PLL applications the VCO frequency is always mixed with a crystal-controlled frequency to yield a difference signal. This difference signal is fed to the programmable divider input. In the vast majority of PLL applications, the VCO frequency is higher than the frequency of the crystal-controlled signal with which it is mixed. With this arrangement, if the VCO frequency increases,

Table 5-8. Midland 77-830 PLL Synthesizer Specifications.

Channel	Programming Input Code Pins							Frequency At TP4		Divisor "N"	Freq/Prog Div./Input MHz
	9	10	11	12	13	14	15	MHz	VCO		
1	1	0	0	1	0	1	0	37.660	17.18	330	3.30
2	1	0	0	1	0	0	1	37.670	17.19	329	3.29
3	1	0	0	1	0	0	0	37.680	17.20	328	3.28
4	1	0	0	0	1	1	0	37.700	17.22	326	3.26
5	1	0	0	0	1	0	1	37.710	17.23	325	3.25
6	1	0	0	0	1	0	0	37.720	17.24	324	3.24
7	1	0	0	0	0	1	1	37.730	17.25	323	3.23
8	1	0	0	0	0	0	1	37.750	17.27	321	3.21
9	1	0	0	0	0	0	0	37.760	17.28	320	3.20
10	0	1	1	1	1	1	1	37.770	17.29	319	3.19
11	0	1	1	1	1	1	0	37.780	17.30	318	3.18
12	0	1	1	1	1	0	0	37.800	17.32	316	3.16
13	0	1	1	1	0	1	1	37.810	17.33	315	3.15
14	0	1	1	1	0	1	0	37.820	17.34	314	3.14
15	0	1	1	1	0	0	1	37.830	17.35	313	3.13
16	0	1	1	0	1	1	1	37.850	17.37	311	3.11
17	0	1	1	0	1	1	0	37.860	17.38	310	3.10
18	0	1	1	0	1	0	1	37.870	17.39	309	3.09
19	0	1	1	0	1	0	0	37.880	17.40	308	3.08
20	0	1	1	0	0	1	0	37.900	17.42	306	3.06
21	0	1	1	0	0	0	1	37.910	17.43	305	3.05
22	0	1	1	0	0	0	0	37.920	17.44	304	3.04
23	0	1	0	1	1	0	1	37.950	17.47	301	3.01
24	0	1	0	1	1	1	1	37.930	17.45	303	3.03
25	0	1	0	1	1	1	0	37.940	17.46	302	3.02
26	0	1	0	1	1	0	0	37.960	17.48	300	3.00
27	0	1	0	1	0	1	1	37.970	17.49	299	2.99
28	0	1	0	1	0	1	0	37.980	17.50	298	2.98
29	0	1	0	1	0	0	1	37.990	17.51	297	2.97
30	0	1	0	1	0	0	0	38.000	17.52	296	2.96
31	0	1	0	0	1	1	1	38.010	17.53	295	2.95
32	0	1	0	0	1	1	0	38.020	17.54	294	2.94
33	0	1	0	0	1	0	1	38.030	17.55	293	2.93
34	0	1	0	0	1	0	0	38.040	17.56	292	2.92
35	0	1	0	0	0	1	1	38.050	17.57	291	2.91
36	0	1	0	0	0	1	0	38.060	17.58	290	2.90
37	0	1	0	0	0	0	1	38.070	17.59	289	2.89
38	0	1	0	0	0	0	0	38.080	17.60	288	2.88
39	0	0	1	1	1	1	1	38.090	17.61	287	2.87
40	0	0	1	1	1	1	0	38.100	17.62	286	2.86
Pin Weight	64	32	16	8	4	2	1				

the input frequency to the programmable divider also increases, The arrangement used in the Midland 77-830 CB PLL circuit works in reverse. The VCO frequency is lower than the signal with which it is mixed to yield the programmable divider input signal. This is shown by the formula:

$$fPDI = 20.480\ MHz - fVCO$$

where fPDI is the input to the programmable divider, and fVCO is the VCO frequency. Therefore, if the VCO frequency increases, the input to the programmable divider will decrease. Stated another way, the frequency at the input of the programmable divider varies inversely as the VCO frequency. In most other PLL arrangements, the frequency at the input of the programmable divider varies directly as the VCO frequency.

The block diagram in Fig. 5-8 more clearly illustrates the comparison between the Midland 77-830 CB synthesizer and the more conventional arrangement. The Midland 77-830 CB synthesizer is shown in Fig. 5-8A, while the more conventional arrangement is shown in Fig. 5-8B. Only the relevant parts of the two synthesizers are shown. The first thing you might notice is that the varactors are hooked up oppositely. In the Midland synthesizer in Fig. 5-8A, the voltage at test point A (TPA) will decrease as the channel selector is switched from channel 1 through channel 40. In the conventional arrangement in Fig. 5-8B, the voltage at test point A increases as the channel selector is switched from channel 1 through channel 40. In both instances, however, the reverse bias on the varactor diode increases as the channel selector is switched from channel 1 through channel 40. This increased reverse bias on the varactor diode increases the VCO frequency in both cases. This is always true in all VCOs.

The VCO frequency and the mixer output frequency are listed for both channels, 1 and 40, in Fig. 5-8. The VCO frequency increases in both synthesizers as the channel increases from 1 through 40. Notice, though, that the output frequency of the mixer decreases in the Midland circuit as the channel increases from 1 through 40. In the conventional circuit of Fig. 5-8B, the mixer output frequency increases as the channel increases from 1 through 40.

Despite this synthesizer working in reverse to most other CB PLL synthesizers, the end result is the same. It is a well designed circuit and is quite reliable.

Let's get back to the discussion on the PLL IC stage, IC1, in the schematic of Fig. 5-6. You can see on the schematic that the

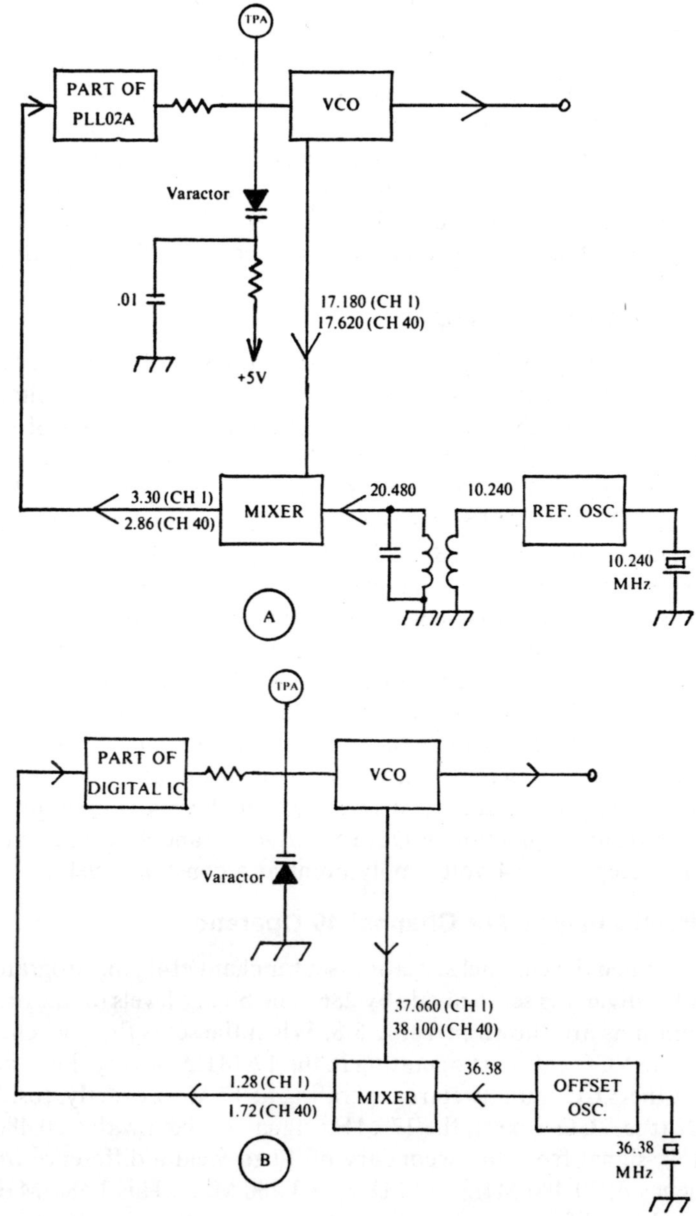

Fig. 5-8. The Loop arrangement used in the Midland 77-830 PLL circuit is shown at A, and a conventional circuit is shown at B.

offset frequency from pin 9 of the VCO and mixer (IC2) is fed through capacitor C20 to the input of the programmable divider. There it is divided by the appropriate divisor N to yield a 10-kHz signal. Pin 3 of IC1 is the input to the fixed reference divider that divides by 1024. The reference oscillator signal (10.240 MHz) is fed to pin 3 of IC1 from the emitter of buffer Q2 through C61. the phase detector output appears at pin 5. This output is filtered by an RC network low-pass filter and is then used to control the VCO. As mentioned before, pin 6 is the lock detector output. The operation of the lock detector was described in the "Transmit Stop Circuit" section.

Regulated 5-Volt Supply

The supply voltage to the digital PLL synthesizer IC must be fairly well regulated. For this reason, the PLL IC supply voltage is usually furnished by a special regulator which is part of the power supply. Figure 5-9 shows the 5.4-volt regulator used to power the PLL02A. Transistor Q6 is a series pass transistor. The base of Q6 is set at a fixed bias by the 9.1-volt zener diode, D3. The voltage at the emitter is 8.7 volts. If the emitter voltage drops, this forward bias on Q6 will increase, causing the conduction to increase. This increased conduction increases the voltage at the emitter, thus compensating for the increased load. If the voltage at the emitter rises, the forward bias on Q6 will decrease, causing the conduction from emitter to collector to decrease. This decreased conduction causes the voltage at the emitter to decrease, thus compensating for the decreased load. The emitter itself serves as one supply voltage point directly, but the 5.4-volt supply is taken from the junction of D2 (a 5-volt zener) and R68. The 5-volt zener keeps the 5.4-volt supply point at a constant level.

Circuit Analysis For Channel 40 Operation

When the channel selector is set for channel 40, the programmable divider is set to divide by 286. The binary levels on the program pins are shown in Table 5-8. When the set is first switched on, the VCO will start operating in the 17-MHz vicinity. Let's say that the VCO initially starts operating at 17 MHz exactly. Inside IC2 (the VCO mixer), the 17-MHz signal is mixed with a 20.480-MHz signal from the secondary of T1 to yield a difference frequency of 20.480 MHz - 17 MHz, or 3.480 MHz. This 3.480-MHz signal is fed from pin 9 of IC2 to the input of the programmable divider at pin 2 of IC1. The divider divides the 3.480-MHz signal by 286 to yield 12.168 kHz at the output of the programmable di-

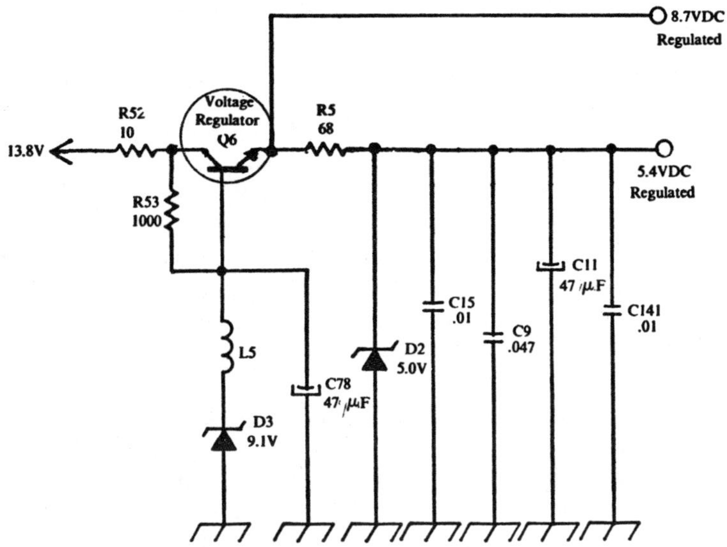

Fig. 5-9. Part of the power supply. The source of the regulated5-volt supply is shown. This 5-volt supply powers the PLL IC.

vider. This 12.168-kHz signal is then compared with the 10-kHz reference frequency in the frequency/phase comparator or detector.

Because a large difference exists between the two frequencies, the frequency/phase detector will produce a large error signal on its output. This error signal appears at pin 5 of IC1. The error signal is filtered by the RC filter network and then applied to the VCO control line. The error signal increases the reverse bias on the varactor diode, D1. Thus, the capacitance of the varactor decreases. As the capacitance of the varactor decreases, the frequency of the VCO will increase. As the VCO frequency increases, the frequency at the input to the programmable divider decreases toward 2.86 MHz and the output of the programmable divider decreases towards 10 kHz. As this happens, the error signal at the output of the frequency/phase detector becomes smaller and smaller. When the VCO frequency reaches 17.620 MHz, the input to the programmable divider will be 20.480 MHz - 17.620 MHz, or 2.86 MHz. As this point, the output of the programmable divider will be 10 kHz, and the VCO control voltage will contain no error signal component. The control voltage stabilizes the VCO fre quency at 17.620 MHz, and the feedback loop keeps it locked at this frequency. This 17.620-Mhz VCO signal is mixed with the 20.480-MHz signal from the secondary of T1 to yield a sum frequency of 38.100 MHz at pin 6 of IC2.

In the receive mode, this 38.100-MHz signal is fed to the first mixer. When mixed with a channel 40 frequency of 27.405 MHz, the mixer will yield a difference frequency of 38.100 MHz - 27.405 MHz, or 10.695 MHz. Thus the first intermediate frequency of 10.695 MHz is developed. This 10.695-MHz signal (1st i-f) is fed to a second mixer. Here, it is mixed with a 10.240-MHz signal from the reference oscillator to yield a difference frequency of 10.695 MHz - 10.240 MHz, or 455 kHz. This 455 kHz signal is the second i-f.

In the transmit mode, the 38.100-MHz signal at pin 6 of IC2 is fed through C23 to pin 4 of IC3. Inside IC3, this 38.100-MHz signal is mixed with a 10.695-MHz signal from the transmit oscillator (also part of IC3). The difference of these two frequencies is taken from pin 9 of IC3. This difference frequency is 38.100 MHz - 10.695 MHz, or 27.405 MHz. This is the transmit frequency. This signal is then sent to the following transmitter stages for further amplification.

Alignment

☐ Connect a frequency counter to TP1 (the emitter of the reference oscillator, Q1) and adjust trimmer capacitor CT1 for exactly 10.240 MHz.

☐ Connect an rf voltmeter to TP2 and tune transformer T1 for maximum rf.

☐ Connect a DC voltmeter to TP8 and with the channel selector on channel 19, adjust L1 for approximately 3.0 volts. This voltage typically ranges from 3.7 volts on channel 1 to 1.9 volts on channel 40.

☐ Connect an rf voltmeter to TP4 and adjust L2 and T2 for maximum rf on channel 19 or 20.

☐ Connect a frequency counter to TP3 and check for 10.695 MHz in the transmit mode (any channel).

☐ Connect an rf voltmeter to the collector of Q3 (the transmit buffer). In the transmit mode, peak tune L5 and T3 for maximum rf voltage.

☐ Connect a frequency counter to TP4 and check for proper frequency on each channel. See Table 5-8 for the proper frequency for each channel.

Troubleshooting

First, what do you do when the rig neither transmits nor receives on any channels? Let's say that you have narrowed the trouble down to the PLL circuit. You should have already measured the signal at TP4 and found it to be missing or of incorrect frequency. As with any electronic circuit, it is good practice to first check the supply voltage. It only takes a moment to do this. A convenient place to check the PLL supply voltage is at pin 1 of IC1. The voltage there should be approximately 5.4 volts. If there is much deviation from this, troubleshoot the regulated 5-volt supply.

Pins 5 and 6 of the PLL02A are very important troubleshooting points. Pin 6 is the lock-detector output, and pin 5 is the phase detector output. If the voltage at pin 6 is approximately 4.5 volts, this is an indication that the PLL loop is in a locked condition. Using the following procedure, you can verify that the loop is locked. Temporarily connect a 0.01-muF capacitor between pin 2 and ground. The voltage level at pin 6 should drop to a LOW level with the capacitor connected. If it does, this eliminates suspecting much of the PLL circuitry — the PLL02A, the VCO, the reference oscillator, and at least part of the VCO and mixer (IC2) must be working properly. If the above test proved the loop to be locked and still the signal at TP4 is not correct, check for the proper signal at pin 6 of IC2. If rf is missing at pin 6, IC2 might be defective or L2 might be mistuned. Check the voltage at pin 6 of IC2. It should be approximately 8.4 volts. If the rf voltage is present at pin 6 of IC2, check capacitor C22 and the tuning of T2 and L2.

If the voltage at pin 6 of IC1 was LOW when you first checked there, then the loop is apparrently in an out-of-lock condition. This can be caused by a number of things. First check for the presence of the reference oscillator signal at pin 3 of IC1. Use your rf voltmeter first. If the rf voltmeter indicates a signal there, then use your counter to check the frequency of the signal. It should be 10.240 MHz. If the reference signal is present and on frequency, go to the following step. If the reference signal ismissing or is far off frequency, troubleshoot the reference oscillator or buffer.

Using an rf voltmeter, check for the presence of a signal at pin 2 of IC1, the programmable divider input. If the signal is missing, troubleshoot the VCO and mixer (IC2). If the signal is present, it is off frequency; otherwise, the loop would be locked and you would not have gotten this far in the troubleshooting! Check the control voltage at pin 5 of IC1. It is probably abnormally HIGH or abnormally LOW. Here are the remaining possibilities: There may be a defect in IC2 (the VCO and mixer). The VCO coil L1 could be drastically mistuned. Also, if the 20.480-MHz signal is missing at pin 4 of IC2, the result would be the same. Check IC2 voltages and the components around it. You might also try tuning L1 while monitoring the voltage at pin 5 of IC1. If the voltage starts changing while you are tuning L1, you're on the right track. The voltage should be set at about 3 volts on channel 19. If tuning L1 does not cause the voltage at pin 5 of IC1 to change, you will need to return the slug in L1 to its original setting.

Wrong Receive and Transmit Frequencies on High Channels with Low Channels OK. These wrong frequencies can be measured by using a counter at TP4. This is usually caused by VCO misalignment. The transmitter will not produce any output because the lock detector stops it. See the alignment instructions for tuning the VCO. The voltage at pin 5 of IC1 will probably be abnormally low if the problem is VCO misalignment.

Wrong Receive and Transmit Frequencies on Low Channels with High Channels OK. This is the same problem as above. This time though, the VCO is probably misadjusted in the opposite direction. The voltage at pin 5 of IC1 will probably be abnormally high. See the VCO alignment instructions.

No Transmit with Receive OK. Check for 10.695 MHz at TP3. Check for proper operation of IC3 (transmit mixer and oscillator).

Transmit Signal Off Frequency with Receive OK. Check for proper frequency at TP3. The 10.695-MHz crystal might have drifted off frequency or a defective component around the oscillator could cause the oscillator to operate off frequency.

Some Channels Inoperative. Check for proper binary levels on the programming pins of IC1. Check the program lines for shorts or opens, as well as the channel switch itself. A defective IC1 could possibly cause this problem also.

Some Channels Intermittent. Check the channel selector for loose or dirty contacts. Clean or replace as necessary.

Chapter 6

The MM55104N PLL IC

The MM55104N is manufactured by National Semiconductor Corporation. This is a CMOS device that is housed in a 16 pin dual-in-line package. The IC contains a reference oscillator (with external crystal), reference divider, programmable divider, and phase detector. The pin identification diagram is shown in Fig. 6-1. Pin number 1 is nearest the indentation. The absolute maximum ratings of the MM55104N are listed in Table 6-1. Notice the wide operating temperature range (-30 Degrees C to + 75 Degrees C). Other electrical characteristics are listed in Table 6-2. A discussion of each internal stage of the MM55104N follows.

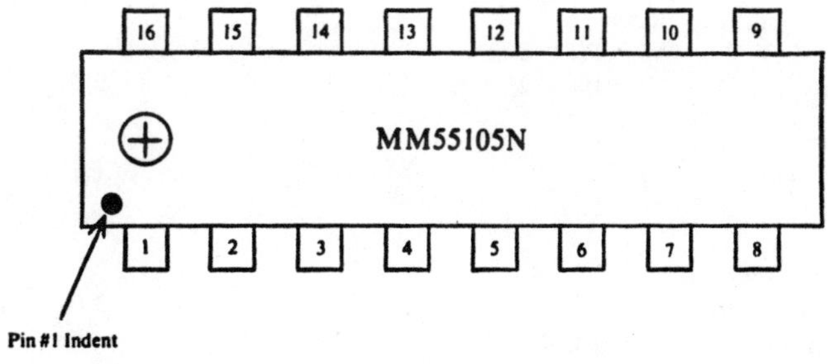

Fig. 6-1. An overview of the MM55104N PLL IC, showing the pin numbering. (Courtesy of National Semiconductor Corp.).

Table 6-1. Absolute Maximum Ratings of the MM55104N (courtesy of National Semiconductor Corp.).

Voltage at any pin ....VCC + 0.3 V to Gnd - 0.3 V
VCC Maximum .... 7 V
Operating temperature range -30 Degrees C to +75 Degrees C
Storage temperature ...... -40 Degrees C to +125 Degrees C
Lead Temperature (soldering, 10 seconds) 300 Degrees C

Table 6-2. Electrical Characteristics of the MM55104N (courtesy of National Semiconductor Corp.).

PARAMETER	TA(Ambient Temperature) = 75 Degrees C CONDITIONS	MIN	TYP	MAX	UNITs
Supply voltage (VCC)		4.5	5.0	5.5	V
Supply current (ICC)	Freq. @ Osc In = 10 MHz @ Fin=2.5 MHz, All other I/O pins open. VCC=5 V		3	10	mA
Logical "1" input voltage P0 - P8, FS, Fin		VCC-0.4			V
Logical "0" input voltage P0 - P8, FS, Fin				0.4	V
Logical "1" output voltage					
O VCO,	IO = 0.4 mA				
LD	IO = 0.5 mA	VCC-0.5			V
Osc Out	IO = 0.25 mA				
Locial "0" output voltage					
O VCO, LD	IO = .5 mA				
Osc Out	IO = .25 mA			0.5	V
Logical "1" input current FS (Pull-up)				1.0	uA
P0 - P7 (Pull-down)	VCC = 5 V	5	20	50	uA
Locial "0" input current P0 - P8 (Pull-down)				1.0	uA
FS (Pull-up)	VCC = 5 V	-10	-35	-100	uA
Max Toggle Freq @ Fin		3			MHz
Max Osc Freq. @ Osc In		10.240			MHz
TRI-STATE Leakage @ O VCO				1.0	uA

REFERENCE OSCILLATOR

This is a CMOS on-chip oscillator with its input connected to terminal pin 3 and its output connected to terminal pin 4. Also, an internal connection is made from the oscillator output to the reference divider input. The crystal which determines the resonant frequency of the oscillator connects between the input and output of the oscillator. The crystal is resonant at 10.240 MHz.

REFERENCE DIVIDER

This divider divides the reference oscillator frequency by 1024 or 2048, depending upon the binary level at pin 5. If the level at pin 5 is binary 1, the divider is set to divide by 1024. This gives a reference frequency of 10 kHz at its output. If the level at pin 5 is binary 0, the divider is set to divide by 2048. This gives a reference frequency of 5 kHz at its output.

There is a pullup resistor connected internally between pin 5 and VCO. If pin 5 is left open or ungrounded, this pullup resistor automatically places pin 5 at binary 1, which gives a 10-kHz reference frequency. To get a 5-kHz reference frequency, pin 5 must be grounded.

PROGRAMMABLE DIVIDER

According to Table 6-2, the maximum guaranteed toggle frequency of the programmable divider is 3 MHz. This means that in order to ensure proper frequency division, the input frequency at pin 2 must be kept well below this limit. Because, the VCOs in all CB PLL systems operate at much higher frequencies than this, the VCO frequency must be down converted before being applied to the programmable divider input.

The programmable divider has eight program lines. The programming to this divider is done with the pure binary code. The maximum divisor which can be programmed into the divider is 255. Each program line has a pulldown resistor connected to it. If a program input is left open, the pulldown resistor automatically places the line at binary 0 or ground. This simplifies the design of the channel selector switch. the switch only has to place binary 1 on the proper pins, the others being automatically placed at binary 0 by the pulldown resistors. Table 6-3 shows the programming code for a couple of divisors, N.

PHASE DETECTOR

The phase detector has an output that can be one of three different states:

□ Provides high level voltage when the VCO frequency is lower than the lock frequency.

□ Provides low level voltage when the VCO frequency is higher than the lock frequency.

□ Provides high impedance (TRI-STATE®) output under lock conditions. Associated with the phase detector is the lock detector. Under normal locked conditions, the lock detector output is at a high voltage level (4.5 volts). If an unlocked condition occurs, the output of the lock detector goes to a low voltage level (not to exceed 0.5 volts).

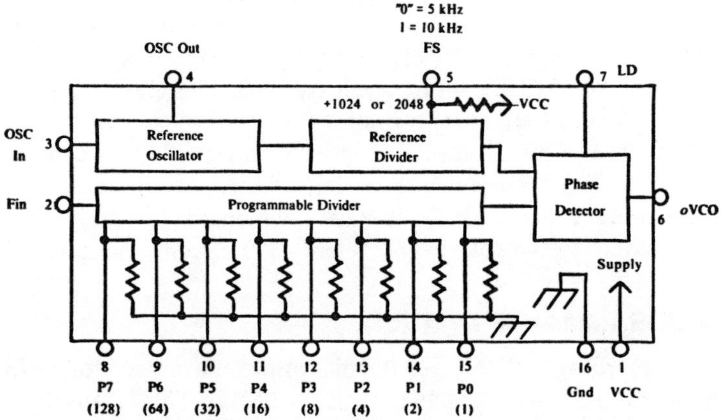

Pin Descriptions
P0 - P7 Programmable divider inputs
Fin - Frequency input from VCO (offset mixer)
Osc In - Oscillator amplifier input terminal
Osc Out - Oscillator output terminal
LD - Lock detector
ØVCO - Output of phase detector for VCO control
FS - Frequency division select 10 or 5 kHz - "1" is 10 kHz "0" is 5 kHz

Fig. 6-2. Internal structure of the MM55104N. Notice the pulldown resistors on the various program pins (8 through 15) and the pullup resistor at pin 5 (FS). The weights of the various program pins are shown in parentheses below the pin number (courtesy of National Semiconductor Corp.).

PRACTICAL APPLICATION

Figure 6-3 shows a block diagram of a practical 23-channel CB PLL synthesizer utilizing the MM55104N IC. This PLL synthesizer requires the use of three crystals, one for the 10.240-MHz reference oscillator, one for the offset oscillator and one for the transmit oscillator.

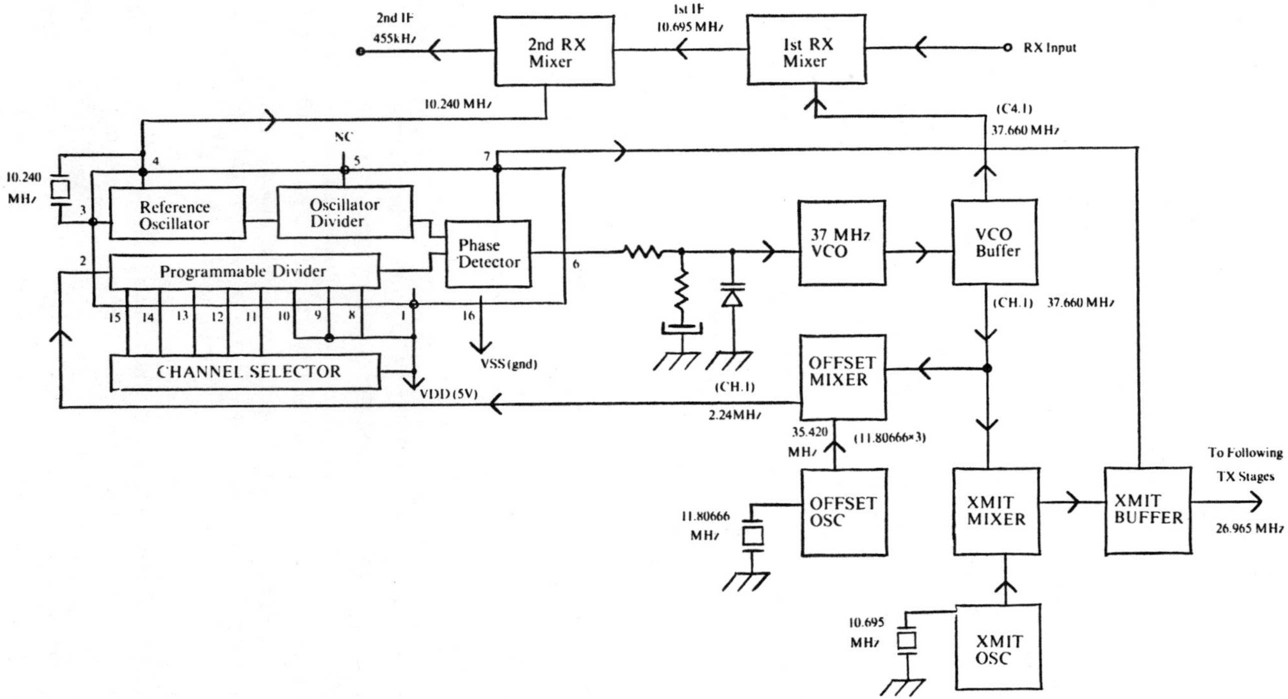

Fig. 6-3. Block diagram of a typical application of the MM55104N. This particular application was used in 23-channel CB sets.

The reference frequency select terminal (pin 5) of the MM55105N is left open. The on-chip pullup resistor automatically places pin 5 at binary 1 level. This sets the reference frequency at 10 kHz. Now check the program lines to see how they are set up.

The program lines are pins 8 through 15. Pin 8, with a weight of 128, is the *most* significant. Pin 15, with a weight of 1, is the *least* significant. Pins 8, 9, and 10 are connected to + 5 volts (VDD) directly, so these pins are kept at binary 1 level at all times. The weight of pins 8, 9, and 10 are 128, 64, and 32, respectively. This sets the minimum divisor N at 128 + 64 + 32, or 224. The other five program lines (11 through 15) are connected to a channel selector switch that can either connect the line to + 5 volts (binary 1) or leave it open (binary 0), depending upon the position of the channel switch.

For 23-channel operation, the N divisor ranges from 224 for channel 1 to 253 for channel 23. The VCO frequency ranges from 37.660 MHz for channel 1 to 37.950 MHz for channel 23. The offset frequency ranges from 2.24 MHz for channel 1 to 2.53 MHz for channel 23. The offset oscillator operates at the third overtone of the 11.80666-MHz crystal to give a frequency of 3 × 11.80666, or 35.420 MHz, at its output. This 35.420-MHz oscillator signal is fed to the offset mixer where it is heterodyned against the 37-MHz VCO frequency to yield a difference frequency at the output of the offset mixer. This offset frequency varies directly as the VCO frequency. If the VCO frequency increases, the offset frequency increases. Conversely, if the VCO frequency decreases, the offset frequency decreases.

The transmit oscillator at 10.695 MHz feeds the transmit mixer along with the 37 MHz VCO signal. The difference of these two frequencies appears at the output of the transmit mixer. This is the actual transmit frequency. This transmit signal is then fed on to the transmit buffer. As mentioned before, pin 7 is the lock detector output. Under normal locked conditions, the voltage at pin 7 is approximately 4.5 volts. This is used to provide forward bias to the transmit buffer stage. If an unlocked condition occurs, the voltage at pin 7 drops to a low level (0.5 volts or less). This action then kills the forward bias to the transmit buffer, killing the transmitter. This in turn prevents off-frequency signal radiation.

Using channel 1 as an example, here is now this PLL operates. For channel 1, the programmable divider is set to divide by 224. If the output of the programmable divider is not exactly 10 kHz, the phase detector produces an output that causes the VCO to move until the frequency at the output of the programmable divider is exactly 10 kHz. For a 10-kHz output, the input to the programmable divider will have to be 10 kHz × 224, or 2.24 MHz. In order for the output of the offset mixer to be 2.24 MHz, the VCO frequency will have to be 2.24 MHz + 35.420 MHz, or 37.660 MHz. This is the proper VCO frequency for channel 1.

In the receive mode, this 37.660-MHz VCO signal is fed to the receiver first mixer. When mixed with a channel 1 frequency of 26.965 MHz, a difference frequency of 10.695 MHz is developed at the output of the first mixer. This is the first intermediate frequency. This 10.695-MHz first intermediate frequency is then fed to a second receiver mixer, where it is mixed with a 10.240-signal from the reference oscillator. The difference frequency appears at the output of the second mixer. This is a 455-kHz second intermediate frequency.

In the transmit mode, the 37.660-MHz VCO frequency is fed to the transmit mixer, where it is heterodyned against a 10.695-MHz signal from the transmit oscillator. The difference of the two frequencies appears at the output of the transmit mixer. This is the transmit frequency for channel 1, 26.965 MHz. Table 6-4 lists the VCO frequency, offset frequency, divisor N, and programming code for each of the 23 channels.

Table 6-3. Decimal Number Versus Programming Code for the MM55104N.

	(128)	(64)	(32)	(16)	(8)	(4)	(2)	(1)
N	P7	P6	P5	P4	P3	P2	P1	P0
1	0	0	0	0	0	0	0	X
2	0	0	0	0	0	0	1	0
255	1	1	1	1	1	1	1	1

Fout = Fin/ N "1" = 4.5 V

"0" = less than 0.4 V

X = Don't care

PEARCE-SIMPSON SUPER TIGER 40A PLL SYNTHESIZER

The block diagram of the Pearce-Simpson Super Tiger 40A PLL synthesizer is shown in Fig. 6-4. The complete schematic of the PLL synthesizer is shown in Fig. 6-5. After each stage of this PLL circuit is discussed, a complete circuit operation and mathematical analysis will be presented using channel 40 as an example. But first the individual stages will be covered.

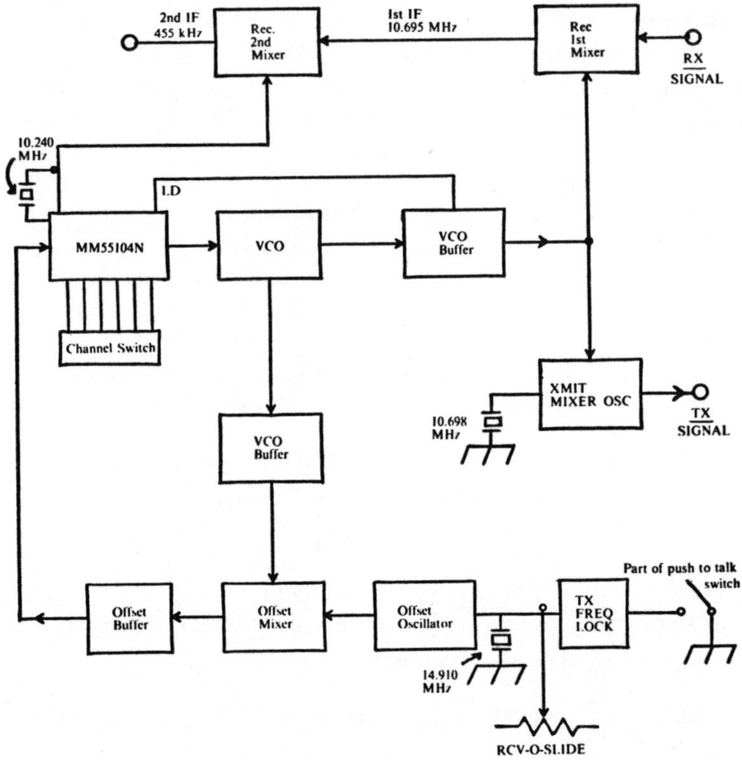

Fig. 6-4. Block diagram of the Pearce-Simpson Super Tiger 40A, which uses the MM55104N. This application requires the use of three crystals.

IC1 - MM55104N

From the schematic shown in Fig. 6-5, you can see that six of the program lines (pins 10 through 15) connect to the channel selector switch. The other two program lines are pins 8 and 9. Pin 8, with a weight of 128, connects directly to 4.5 volts through resistor R77. This keeps pin 8 at the binary 1 level at all times and thus establishes the minimum divisor of 128. Pin 9 (with a weight of 64) is connected to ground through resistor R74. This keeps pin 9 at the binary 0 level at all times so its weight of 64 will never be used in determining the divisor N. The other 6 program lines can be at either the binary 1 or binary 0 level, depending upon the position of the channel selector switch.

The 10.240-MHz reference oscillator crystal (X2) is connected between pins 3 and 4 of IC1. This is the input and output of

Table 6-4. Pearce-Simpson Super Tiger 40A PLL Synthesizer Data.

Channel No.	Programming Pin No. (16) 11	(8) 12	(4) 13	(2) 14	(1) 15	"N" Divisor	VCO FREQ MHz	Offset Freq. MHz
1	0	0	0	0	0	224	37.660	2.24
2	0	0	0	0	1	225	37.670	2.25
3	0	0	0	1	0	226	37.680	2.26
4	0	0	1	0	0	228	37.700	2.28
5	0	0	1	0	1	229	37.710	2.29
6	0	0	1	1	0	230	37.720	2.30
7	0	0	1	1	1	231	37.730	2.31
8	0	1	0	0	1	233	37.750	2.33
9	0	1	0	1	0	234	37.760	2.34
10	0	1	0	1	1	235	37.770	2.35
11	0	1	1	0	0	236	37.780	2.36
12	0	1	1	1	0	238	37.800	2.38
13	0	1	1	1	1	239	37.810	2.39
14	1	0	0	0	0	240	37.820	2.40
15	1	0	0	0	1	241	37.830	2.41
16	1	0	0	1	1	243	37.850	2.43
17	1	0	1	0	0	244	37.860	2.44
18	1	0	1	0	1	245	37.870	2.45
19	1	0	1	1	0	246	37.880	2.46
20	1	1	0	0	0	248	37.900	2.48
21	1	1	0	0	1	249	37.910	2.49
22	1	1	0	1	0	250	37.920	2.50
23	1	1	1	0	1	253	37.950	2.53

Binary 1 = 4.5 volts
Binary 0 = near 0 volts

the reference oscillator. The reference oscillator also serves as a second receiver oscillator. Notice that pin 4 of IC1 connects to the second mixer. The 10.240-MHz oscillator can be set precisely on frequency by adjusting C73.

Notice that pin 5 is left open. This sets the reference frequency at 10 kHz. Pin 7 (the lock detector output) furnishes forward bias to the VCO buffer stage (TR11) through the secondary of T14 and through R73. As long as the loop is locked, pin 7 will furnish 4.5 volts. If the loop becomes unlocked, however, the voltage at pin 7 will fall to near 0 volts, thus killing the forward bias to TR11 which stops the transmitter output. The output from pin 6 is the

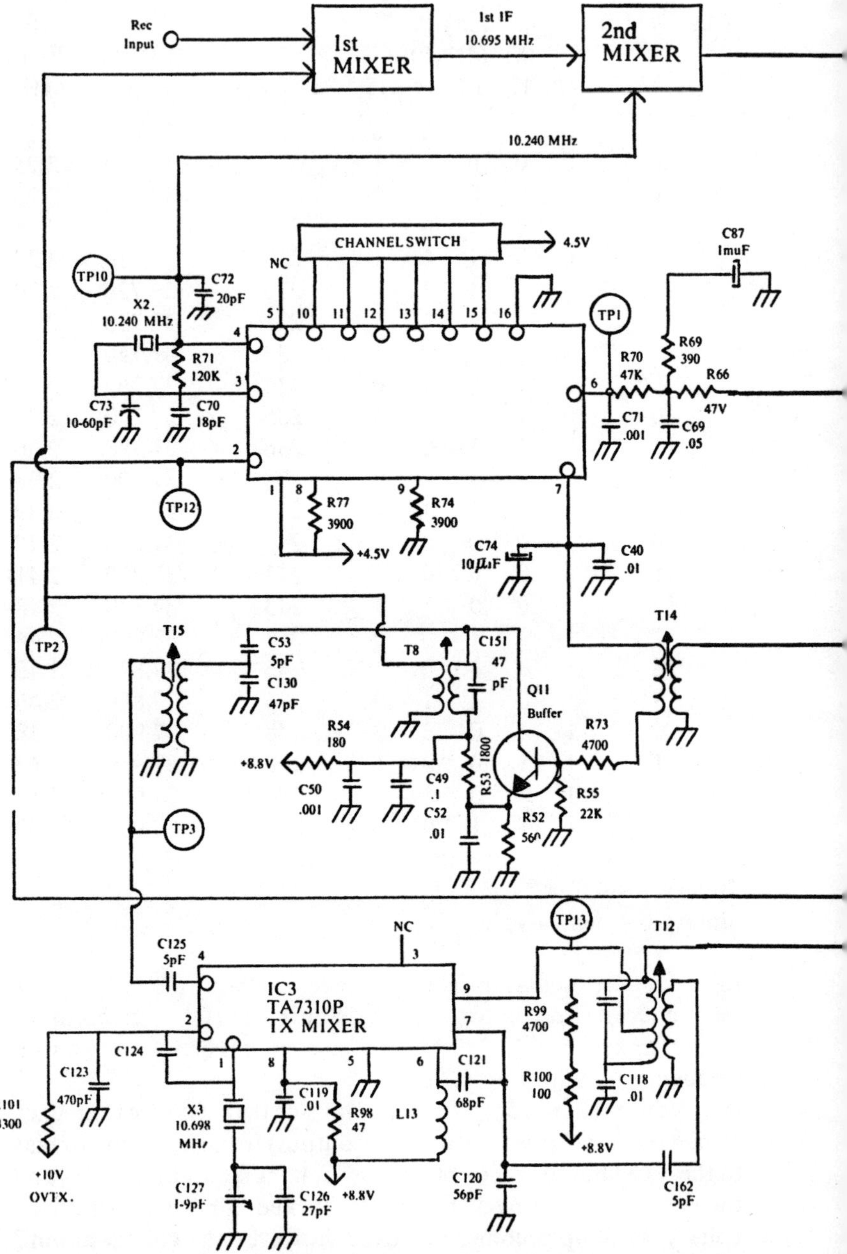

Fig. 6-5. the PLL synthesizer used in the Pearce-Simpson Super Tiger 40A.

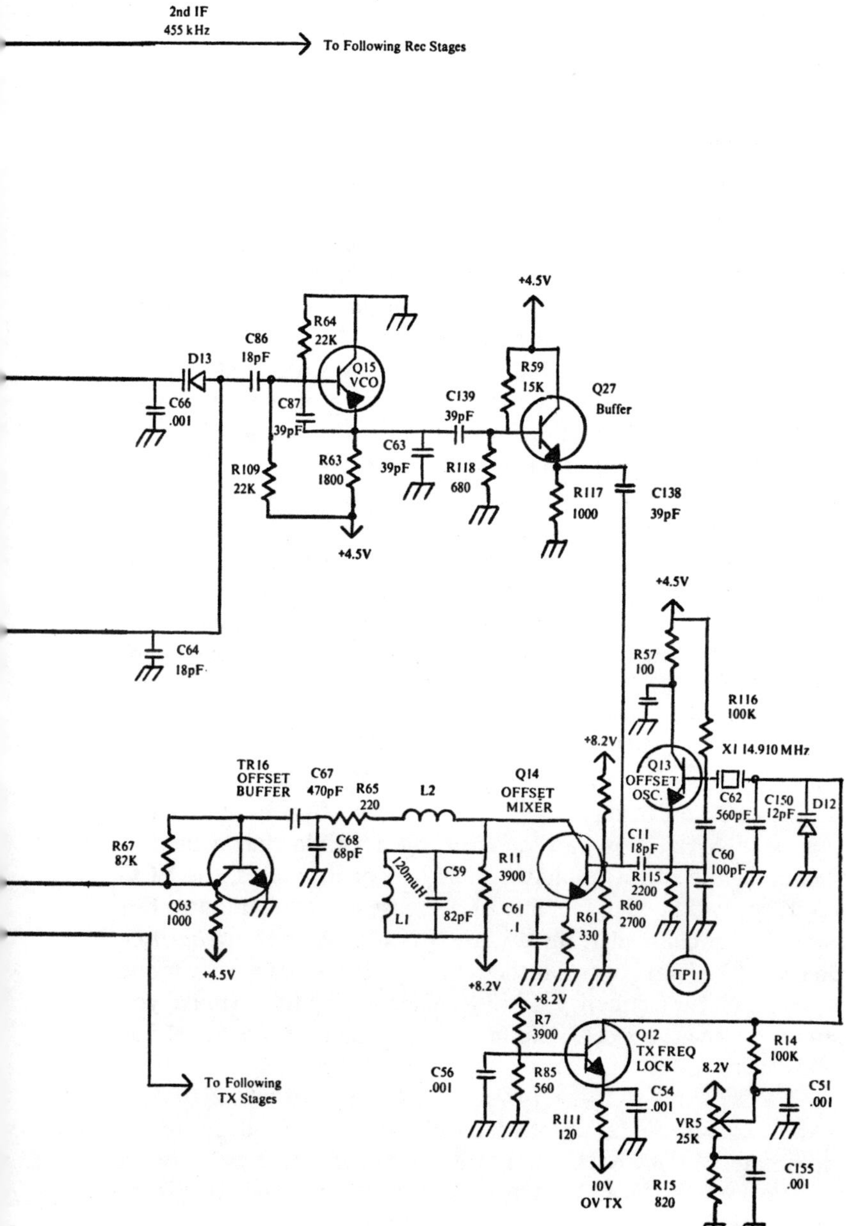
2nd IF
455 kHz
To Following Rec Stages
+4.5V
R64
22K
C86
18pF
D13
Q15
VCO
R59
15K
Q27
Buffer
C66
.001
C87
39pF
C139
39pF
C63
39pF
R63
1800
R118
680
R109
22K
R117
1000
C138
39pF
+4.5V
C64
18pF
R57
100
R116
100K
+8.2V
X1 14.910 MHz
TR16
OFFSET
BUFFER
C67
470pF
R65
220
L2
Q14
OFFSET
MIXER
Q13
OFFSET
OSC.
C62
560pF
C150
12pF
D12
C68
68pF
C11
18pF
R67
87K
120muH
L1
C59
82pF
R11
3900
C61
.1
R115
2200
C60
100pF
Q63
1000
R61
330
R60
2700
TP11
+4.5V
+8.2V
+8.2V
R7
3900
Q12
TX FREQ
LOCK
R14
100K
C56
.001
R85
560
8.2V
To Following
TX Stages
C54
.001
C51
.001
R111
120
VR5
25K
C155
.001
10V
OV TX
R15
820

phase detector output. Notice that this output is filtered by an RC network that forms a low-pass filter. The filtered voltage is then used to set the reverse bias on the varactor diode, D13, to control the VCO frequency.

VCO Stage

The VCO stage consists of transistor TR15 and the varactor diode D13, along with other associated components. Also associated with the VCO stage are two separate VCO buffers, TR11 and TR27. These will be considered part of the VCO stage. The frequency of the VCO is controlled by a positive voltage that is applied to the cathode of the varactor diode, D13. The cathode of the varactor is placed at rf ground potential by C66, a 0.001-muF capacitor. The anode is connected to the top of the primary of transformer T14. The primary of T14 also serves to place the anode of D13 at DC ground potential.

The VCO operates in the 16-MHz range. The VCO operates at different frequencies in the transmit and receive modes. In the transmit mode, the VCO operates at a frequency that is approximately 3 kHz lower than the receive mode frequency. This is because the offset oscillator operates about 3 kHz lower in the transmit mode than it does in the receive mode. The operation of the offset oscillator will be fully discussed next. The VCO frequency can be determined from the following formula:

$$FVCO = N \times 10^{-2} + F \text{ off osc}$$

where FVCO is the VCO frequency in MHz, N is the divisor of programmable divider and F off osc is the offset oscillator frequency in MHz. You can see from the formula that if the frequency of the offset oscillator increases or decreases, the VCO frequency will have to increase or decrease an equal amount. For example, on channel 1, the divisor N is 136. With the *delta-tune (Receive-O-Slide)* control is the center position, the receive mode offset oscillator frequency is 14.910 MHz. Using the formula, you can determine the receive mode. VCO frequency for channel 1 as follows:

$$FVCO = 136 \times 10^{-2} + 14.910 = 1.36 + 14.910 = 16.270 \text{ MHz}$$

In the transmit mode, the offset oscillator frequency drops to approximately 14.907 MHz. In the transmit mode, the new VCO frequency becomes $FVCO = 136 \times 10^{-2} + 14.907 = 1.36 + 14.907$, or 16.267 MHz. You can see from these calculations that a change in the offset oscillator frequency will cause an equal change in the VCO frequency.

The VCO signal is taken from the emitter of TR15 and fed to the base of the buffer transistor TR27 through C139. The buffer transistor TR27 is connected as an emitter follower.

Another sample of the VCO signal is taken from the secondary of T14 and fed to the base of buffer transistor TR11. At the output of TR11, part of the signal is fed through T8 to the 1st receive mixer and another part of the signal is taken from T15 and fed to the transmit mixer.

Table 6-5 lists the VCO frequencies for the transmit and receive modes for each channel. Also listed are the N divisors, offset frequencies, and programming codes.

Offset Oscillator

The offset oscillator (sometimes called *down oscillator*) consists of transistor TR13 and the associated components. Also associated with the offset oscillator are the *transmit frequency lock stage* (TR12) and the *Receive-O-Slide* circuit. The operation of this offset oscillator is a little unusual. Notice that the varactor diode D12 is connected from one side of the 14.910-MHz crystal (X1) to ground. In the receive mode, the varactor receives its bias from the arm of the Receive-O-Slide control (VR5) through R14. With the Receive-O-Slide control in the center position, the offset oscillator operates at 14.910 MHz. If the Receive-O-Slide control is moved to one side of the center position, the reverse bias on the varactor D12 will be reduced, thus causing the oscillator frequency to decrease. If the Receive-O-Slide control is moved to the other side of the center position, the reverse bias on the varactor D12 will be increased, thus causing the oscillator frequency to increase.

According to the specifications on this radio set, the Receive-O-Slide control changes the receive frequency + 1000 Hz from the center frequency. Thus, the offset oscillator frequency must change by + 1000 Hz. This means that the offset oscillator frequency can be anything from 14.909 MHz to 14.911 MHz as the Receive-O-Slide control is varied from one end to the other. In the center position, the offset oscillator should operate at 14.910 MHz. It was explained in the "VCO Stage" section how changing the offset oscillator frequency causes the VCO frequency to shift an equal amount.

In the transmit mode, it is necessary and desirable to disable the Receive-O-Slide control so that it can't affect the transmit frequency. This is done by transistor TR12, which is called the *transmitter frequency lock stage*. In the receive mode, transistor

Table 6-5. VCO Frequencies.

CHANNEL	"N"	Programming Pins:		Code				Offset Freq. MHz	VCO Frequency RX Mode - MHz	VCO Frequency TX Mode - MHz
		10	11	12	13	14	15			
1	136	0	0	1	0	0	0	1.36	16.270	16.267
2	137	0	0	1	0	0	1	1.37	16.280	16.277
3	138	0	0	1	0	1	0	1.38	16.290	16.287
4	140	0	0	1	1	0	0	1.40	16.310	16.307
5	141	0	0	1	1	0	1	1.41	16.320	16.317
6	142	0	0	1	1	1	0	1.42	16.330	16.327
7	143	0	0	1	1	1	1	1.43	16.340	16.337
8	145	0	1	0	0	0	1	1.45	16.360	16.357
9	146	0	1	0	0	1	0	1.46	16.370	16.367
10	147	0	1	0	0	1	1	1.47	16.380	16.377
11	148	0	1	0	1	0	0	1.48	16.390	16.387
12	150	0	1	0	1	1	0	1.50	16.410	16.407
13	151	0	1	0	1	1	1	1.51	16.420	16.417
14	152	0	1	1	0	0	0	1.52	16.430	16.427
15	153	0	1	1	0	0	1	1.53	16.440	16.437
16	155	0	1	1	0	1	1	1.55	16.460	16.457
17	156	0	1	1	1	0	0	1.56	16.470	16.467
18	157	0	1	1	1	0	1	1.57	16.480	16.477
19	158	0	1	1	1	1	0	1.58	16.490	16.487
20	160	1	0	0	0	0	0	1.60	16.510	16.507
21	161	1	0	0	0	0	1	1.61	16.520	16.517
22	162	1	0	0	0	1	0	1.62	16.530	16.527
23	165	1	0	0	1	0	1	1.65	16.560	16.557
24	163	1	0	0	0	1	1	1.63	16.540	16.537
25	164	1	0	0	1	0	0	1.64	16.550	16.547
26	166	1	0	0	1	1	0	1.66	16.570	16.567
27	167	1	0	0	1	1	1	1.67	16.580	16.577
28	168	1	0	1	0	0	0	1.68	16.590	16.587
29	169	1	0	1	0	0	1	1.69	16.600	16.597
30	170	1	0	1	0	1	0	1.70	16.610	16.607
31	171	1	0	1	0	1	1	1.71	16.620	16.617
32	172	1	0	1	1	0	0	1.72	16.630	16.627
33	173	1	0	1	1	0	1	1.73	16.640	16.637
34	174	1	0	1	1	1	0	1.74	16.650	16.647
35	175	1	0	1	1	1	1	1.75	16.660	16.657
36	176	1	1	0	0	0	0	1.76	16.670	16.667
37	177	1	1	0	0	0	1	1.77	16.680	16.677
38	178	1	1	0	0	1	0	1.78	16.690	16.687
39	179	1	1	0	0	1	1	1.79	16.700	16.697
40	180	1	1	0	1	0	0	1.80	16.710	16.707
		32	16	8	4	2	1			

Note: means similar to or approximately

Binary 1 = 4.5 volts
Binary 0 = 0 volts

The various weights of the program pins are shown in parenthesis below the pin numbers at the bottom of the column.

TR12 is turned off. In the transmit mode, however, TR12 is turned on by grounding one end of R111 (the emitter resistor) through the microphone push-to-talk switch. When TR12 turns on, it disables the Receive-O-Slide circuit by placing a low-resistance DC ground return at the top of R14. This DC ground return through the emitter/collector of TR12, the 120-ohm emitter resistor (R111), and the microphone push-to-talk switch. Capacitor C54 (0.001 muF) serves as an rf ground for crystal X1 in the transmit mode. This causes the offset oscillator to operate at approximately 14.907 MHz. Notice that this is approximately 3000 Hz lower than its center frequency in the receive mode.

Offset Mixer

The offset mixer consists of transistor TR14 and the associated components. The function of this mixer is to provide an offset frequency for the programmable divider input. This offset frequency is equal to the VCO frequency - 14.910 MHz (receive mode) or VCO frequency - 14.907 MHz (transmit mode). Notice that one of the inputs to the base comes from the offset oscillator and the other comes from the emitter of the VCO buffer, TR27. In the collector circuit of the offset mixer (TR14), you will notice a resonant circuit consisting of coil L1 and capacitor C59. L1 is a 120-muH coil and C59 is an 82-pF capacitor. The resonant frequency is determined by the formula: $F = \frac{1}{\sqrt{2\pi \quad LC}}$

Substituting we have:

$$\frac{1}{2\pi\sqrt{120\times 10^{-6} \times 82 \times 10^{-12}}} = \frac{1}{2\pi\sqrt{9840 \times 10^{-18}}} =$$

$$\frac{1}{2\pi\sqrt{98.4\times 10^{-16}}} = \frac{1}{2\pi(\sqrt{98.4})(10^{-8})} = \frac{1}{2\pi(9.9)(10^{-8})} = \frac{1}{62.2\times 10^{-8}} =$$

$$\frac{1}{0.622\times 10^{-6}} = 1.6\times 10^{6} = 1.6 \text{ MHz.}$$

This 1.6 MHz is exactly what the offset frequency should be for channel 20, the middle of the CB band. The offset frequency for channel 1 is 1.36 MHz. For channel 40, it is 1.80 MHz. Notice that a 3900-ohm resistor (R11) is connected across the LC resonant circuit. This resistor increases the bandwidth of the resonant circuit so that the lowest and highest offset frequencies are passed without much attenuation. The output of the offset mixer is fed to the offset buffer.

Offset Buffer

The offset buffer is a simple stage consisting of TR16 and resistors R67 and R63. The output from the collector feeds the programmable divider input at pin 2 of IC-1.

Transmit Oscillator/Mixer - IC-3

This stage consists of IC3, a TA7310P, and its associated circuitry. This IC contains an oscillator that is resonated by the 10.698-MHz crystal connected to pin 1. The frequency of this oscillator can be adjusted by trimmer capacitor C127. The VCO signal from transformer T15 is fed into the mixer at pin 4 through a 5-pF coupling capacitor, C125. Inside IC3, the VCO frequency is mixed with the 10.698-MHz oscillator signal. The sum of these two frequencies is taken from the top of T12 and fed to the following transmitter stages.

In the receive mode, +10 volts is applied to pin 2 of IC3 through R101. This reverse biases the 10.698-Mhz oscillator transistor within IC3, thus disabling the oscillator. If this oscillator were allowed to operate in the receive mode, the signal might mix with the 10.695-MHz i-f to produce an undesirable beat note. When the transmitter is keyed, one end of R101 is grounded through the microphone push-to-talk switch. This establishes proper bias on the 10.698-MHz oscillator transistor inside IC3, thus the oscillator is enabled.

The transmit mode VCO frequencies are listed in Table 6-5. It is important to note that the transmit mode VCO frequency might deviate slightly from the exact figure listed in the column. Any deviation should be consistent throughout the 40 channels. For example, if the transmit mode VCO frequency for channel 40 were 1000 Hz lower than the 16.267 MHz listed, the transmit mode VCO frequency for all other channels should also be 1000 Hz lower than the frequency listed. To compensate for this, the 10.698-MHz oscillator is adjusted 1000 Hz higher by tuning C127. Actually, you really don't need to know the exact frequency of the VCO or the 10.698-MHz oscillator as long as the sum of the two adds up to the correct channel frequency. The following formula may better clarify this. fCH = fTO + fVCO, where fCH is the channel frequency in MHz, fTO is the transmitter oscillator frequency in MHz, and fVCO is the transmit mode VCO frequency in MHz. You can see that if fVCO is high, fTO can be lowered to compensate, with the sum (fCH) remaining the same. So adjust C127 for proper on-air frequency and don't worry about the fTO and fVCO frequencies individually.

LED Channel Display

Figure 6-6 shows the wiring of the LED channel display. The display consists of two seven-segment LEDs built on one chip. In the figure, the dark segments are the ones which are energized for channel 40. The switches are shown as individual switches for simplicity. Actually, a separate wafer on the channel selector switch applies voltage to the proper segments for each channel. Each segment has a series current limiting resistor to hold the current to a safe level. Pin 9 of the LED device is common. Because this is connected to ground (negative), this is a common-cathode LED display. A common-anode type could be used, but then the common terminal would connect to the positive supply and the channel switch would ground the various segments.

Figure 6-7 shows a rear view of the LED display. The various pins are numbered as shown.

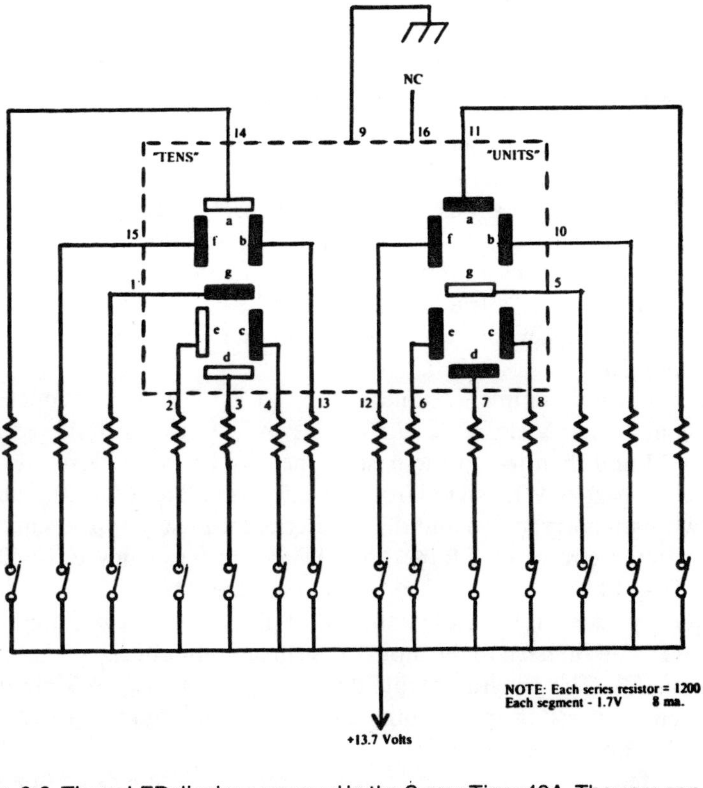

Fig. 6-6. These LED displays are used in the Super Tiger 40A. They are constructed on a single chip. The darkened segments are the ones energized for channel 40.

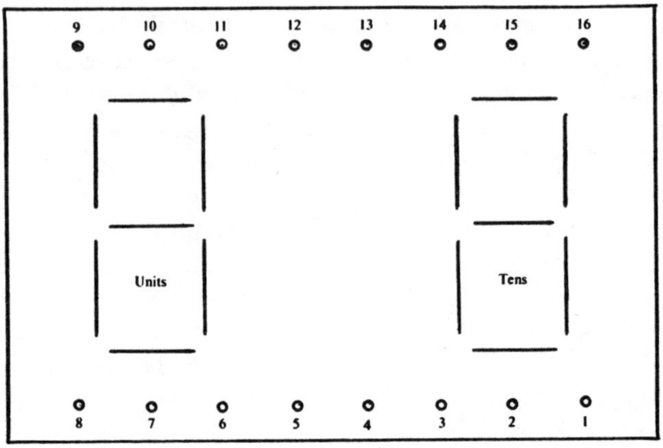

Fig. 6-7. This is a rear view of the LED display. The pin numbering scheme is shown.

Circuit Analysis For Channel 40 Operation

When the set is switched to channel 40, the programmable divider (part of IC1) is set to divide its input frequency by 180. Table 6-5 calls for a VCO frequency of 16.710 MHz. When the set is first switched on, the VCO will initially start operating at some random frequency in the vicinity of 16 to 17 MHz. It takes the PLL action only a fraction of a second to steer the VCO to the proper frequency, but what happens in that fraction of a second is important.

In this example, assume that the VCO initially starts operating at 17.500 MHz. This 17.500-MHz VCO signal is buffered by TR27 and then fed to the base of the offset mixer (TR14) along with a 14.910-MHz signal from the offset oscillator (TR13). These two signals are mixed and the difference frequency appears in the collector circuit of TR14. This difference frequency is 17.500-MHz - 14.910 MHz, or 2.59 MHz. This 2.59 MHz difference frequency (commonly called the offset frequency) is buffered by TR16 and then fed to the input of the programmable divider at pin 2 of IC1. The programmable divider divides this 2.59-MHz frequency by 180 (as programmed). The output from the programmable divider is 2.59 ÷ 180, or 14.4 kHz. The 14.4-kHz signal is then fed to the frequency/phase detector, where it is compared with a 10-kHz reference signal. The frequency/phase detector detects the large frequency difference of its two inputs and issues a proportionately large correction signal at pin 6 of IC1. This correction signal is filtered by the RC filter network and then applied to the VCO control line at the cathode of varactor D13. The resultant correction voltage reduces the positive voltage at the cathode of varactor D13, thus reducing the reverse bias on the varactor. The capacitance of the varactor therefore increases. This in turn lowers the VCO operating frequency.

As the VCO frequency approaches 16.710 MHz, the input to the programmable divider (at pin 2 of IC1) approaches 1.80 MHz, and the output of the programmable divider approaches the 10-kHz reference frequency. As this is progressing, the error signal at pin 6 becomes progressively smaller. When the VCO frequency reaches 16.710 MHz, the input to the programmable divider will be 1.80 MHz and the output of the programmable divider will be 10 kHz (the same as the reference frequency). The control voltage then stabilizes the VCO at 16.710 MHz.

During the time that the PLL was unlocked, the voltage at pin 7 of IC1 (the lock detector output) was at a low level. This disabled the buffer TR11 by killing the forward bias. Because the buffer stage (TR11) is common to both the transmitter and receiver, during an out-of-lock condition both the receive and transmit modes will both be inoperative. When the PLL circuit achieves lockup, the voltage at pin 7 goes high (approximately 4.5 volts). This forward biases transistor TR11 to enable the synthesizer output.

In the receive mode, the 16.710-MHz VCO signal is fed to the receiver first mixer. There, it is heterodyned with the incoming receive signal to yield the first intermediate frequency. For channel 40, the 27.405-MHz signal beats with the 16.710-MHz signal to yield a difference frequency of 27.405 MHz - 16.710 MHz or 10.695 MHz. This 10.695-MHz first i-f signal is fed to the receiver second mixer. A portion of the 10.240-MHz reference oscillator signal (from pin 4 of IC1) also feeds the second mixer. These two signals are heterodyned to yield 10.695 MHz - 10.240 MHz, or 455 kHz. This is the second intermediate frequency.

The Receive-O-Slide control can move the VCO frequency up or down approximately 1000 Hz in the receive mode to compensate for off frequency signals. The Receive-O-Slide control actually varies the *offset oscillator* frequency. When the offset oscillator frequency changes, however, the VCO frequency is forced to change an equal amount in order to keep the offset frequency at pin 2 of IC1 constant. The Receive-O-Slide circuit was discussed thoroughly in the "Offset Oscillator" section.

In the transmit mode, the offset oscillator operates about 3000 Hz lower in frequency (approximately 14.907 MHz). This was also explained earlier. This causes the VCO frequency to shift from 16.710 MHz down to approximately 16.707 MHz. This 16.707-MHz VCO frequency is fed to pin 4 of IC3. Where, it is mixed with a 10.698-MHz signal from the crystal-controlled oscillator that is part of IC3. The sum of these two frequencies appears at the output of IC3 and is taken from T12. This sum frequency is 16.707 MHz + 10.698 MHz, or 27.405 MHz. This is the proper frequency for channel 40. It is important to remember that the VCO frequency shifts approximately 3 kHz between the receive and transmit modes.

Alignment

☐ Set the channel selector in the channel 19 position.

☐ Connect a voltmeter between ground and pin 6 of IC1.

☐ Adjust T14 core to obtain 2.5 volts (+ 0.1 volt) DC.

☐ Rotate the channel switch to channel 1, the reading should be approximately 2.0 volts.

☐ Rotate the channel switch to channel 40, the reading should be approximately 3.0 volts.

☐ Connect an oscilloscope or rf voltmeter to TP3.

☐ Adjust T8 and T15 for maximum rf output on channel 19.

☐ Set the front panel Receive-O-Slide control to midposition.

☐ Set the channel selector to channel 19.

☐ Connect a frequency counter to TP3 (through a 1000-pF coupling capacitor).

☐ Adjust trimmer capacitor C73 to obtain a frequency reading of 16.490 MHz.

☐ Set up a frequency counter to check the transmitter signal.

☐ Set channel selector to channel 19.

☐ Adjust trimmer capacitor C127 for 27.185 MHz with the transmitter keyed.

☐ Check all channels for proper transmit frequency.

General Troubleshooting

PLL synthesizer troubles usually fall into one of two categories: either there is no output from the synthesizer or the output frequency is incorrect. In this synthesizer, the transmit mode synthesizer output appears at TP13 (pin 9 of IC3). The receive mode synthesizer output appears at TP2, the secondary of T8. The following troubleshooting procedures should put you on the right track towards solving PLL troubles, given the various symptoms.

No Receive or Transmit on Any Channel. As always, when everything seems dead, check the supply voltage first. Check for 4.5 to 5.0 volts at pin 1 of IC1. If the voltage is missing, troubleshoot the power supply. If the supply voltage is normal proceed to the next step.

Check the DC voltage level at pin 7 of IC1. This is the *lock detector output.* If the voltage level there is approximately 4.5 volts, proceed to the next step. If the voltage there is low (near 0 volts), check for rf voltage at pin 4 of IC1. If rf voltage is present, use a frequency counter to make certain the frequency is correct (10.240MHz). If rf voltage is missing or the frequency is incorrect, troubleshoot the 10.240-MHz reference oscillator. If the signal at pin 4 is good, check the rf voltage and frequency at pin 2 of IC1. If rf is missing, troubleshoot the VCO, VCO buffer (TR27), offset oscillator, offset mixer, and offset buffer. If rf is present at pin 2 of IC1, but the frequency is wrong, the VCO might be misaligned or a defective varactor might be preventing the phase detector from controlling the VCO frequency. Also, check the VCO control line from pin 6 of IC1 to the cathode of varactor D13. Check for open resistors or leaky capacitors. The voltage at the cathode of varactor D13 should be approximately the same as the voltage at pin 6 of IC1. If the VCO is simply out of alignment, tune L14 for 2.5 volts at pin 6 of IC1 with the channel selector on channel 19. Tune L14 slowly, so you do pin 6 of IC1 with the channel selector on channel 19. Tune L14 slowly, so you do not pass the proper setting. If tuning L14 has no effect on the voltage at pin 6 of IC1, return the core of L14 to its original setting. If all other stages check out OK, consider that IC1 itself might be defective.

If the DC voltage level at pin 7 is normal (4.5 volts), this removes suspicion from all stages connected with the loop. The most likely suspect in this case would be the VCO buffer stage (TR11 and associated circuitry).

Wrong Synthesizer Frequency on High Channels on Transmit and Receive Modes. This is probably caused by improper alignment of the VCO. The voltage at pin 6 of IC1 will probably be higher than normal. Adjust the VCO as explained under the alignment instructions.

Wrong Synthesizer Frequency, Low Channels - Transmit

and Receive Modes. This is also usually caused by VCO misadjustment. This time the VCO is probably misadjusted in the opposite direction, as indicated by an abnormally low voltage reading at pin 6 of IC1. Adjust the VCO as explained in the alignment instructions.

No Transmit with Receive OK. Check the operation of IC3. The 10.698-MHz oscillator may be dead. Check rf voltage at pins 4 and 2 of IC3. IC3 might be defective. Check the various pin voltages.

Wrong Synthesizer Frequencies on Transmit Mode with Receive Mode OK. Check for proper operation of the transmit frequency lock circuit (TR12). Also check the frequency of the 10.698-MHz oscillator. Adjust C127 if incorrect.

Some Channels Inoperative. Check for proper binary levels on the program lines to IC1. See Table 6-5 for the proper levels. Check for a defective channel selector switch. IC1 might be defective.

Some Channels Intermittent. Check for loose or dirty channel selector contacts. Repair or clean as necessary.

Chapter 7

The muPD858C PLL IC

The muPD858C is a custom made IC. It is manufactured by NEC for Uniden, which manufacturers CB radios for several major brands. The device is a CMOS, 24-pin, dual-in-line pin IC. It is used in several major brand CB sets, including the one described in this chapter. The IC contains an on-chip amplifier for use as a reference oscillator, reference frequency divider, frequency/phase detector-charge pump, and an on-chip amplifier that is used as an active filter with external RC components. The pin identification diagram is shown in Fig. 7-1.

SPECIAL FEATURES

- ☐ Single 5-volt supply
- ☐ Low-power CMOS
- ☐ BCD programming
- ☐ Selectable reference frequency (5 kHz or 10 kHz)
- ☐ On-chip oscillator (controlled by external crystal)
- ☐ On-chip filter amplifier (external components)
- ☐ 5.12-MHz output
- ☐ Lock detector output
- ☐ 24-pin dual-in-line package
- ☐ Test points brought out to terminals

INTERNAL STRUCTURE

The block diagram in Fig. 7-2 shows the internal structure of the muPD858C. Let's now check the function of each pin of the IC in numerical order from pins 1 through 24. Pin 1 is the lock detec-

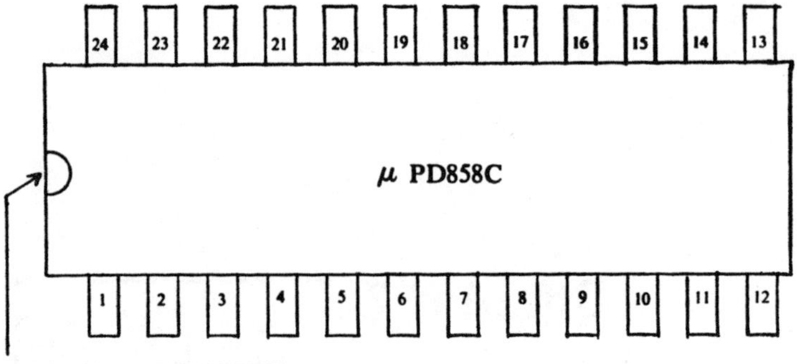

Fig. 7-1. An overview of the muPD858C, showing the pin identification.

tor output, labeled UI (for unlock indicator) in Fig. 7-2. In normal, locked-loop conditions, the voltage at pin 1 is at a LOW level (near 0 volts). If the loop becomes unlocked, the voltage at pin 1 goes to a HIGH level. This high voltage level can be used directly or indirectly to kill the transmitter output. Pin 2 is the output of the charge pump. This is the error signal which will be used to correct the VCO frequency. This signal must be properly filtered before being applied to the VCO control line.

Pin 3 is the input to an on-chip amplifier, which when used with external RC compnents, serves as an active low-pass filter for the charge pump output signal. Pin 4 is the output of the active filter. Figure 7-2 shows that the output of the charge pump at pin 2 is connected to the input of the active filter at pin 3. Pin 5 is the reference input to the phase detector/charge pump. Pin 6 is the output of the reference divider. Usually, the connection from the output of the reference divider to the input of the phase detector is made internally in the IC. In this IC, the connections are made through an external connection between pins 5 and 6. This makes it very convenient to check the output frequency of the reference divider, simply by connecting a frequency counter to pin 6.

Pin 7 (labeled M in Fig. 7-2) is the reference frequency select pin. If a high level voltage (VDD) is placed on pin 7 the reference divider will divide by 1024 to yield a 10-kHz reference signal. If pin 7 is connected to ground (VSS), the reference divider will divide by 2048 to yield a 5-kHz reference signal. Pin 8 is a 5.12-MHz output (10.240 MHz ÷ 2). This is useful in certain CB applications, usually to reduce the number of crystals required. Pin 9 is the input to an on-chip amplifier that is usually used as the reference

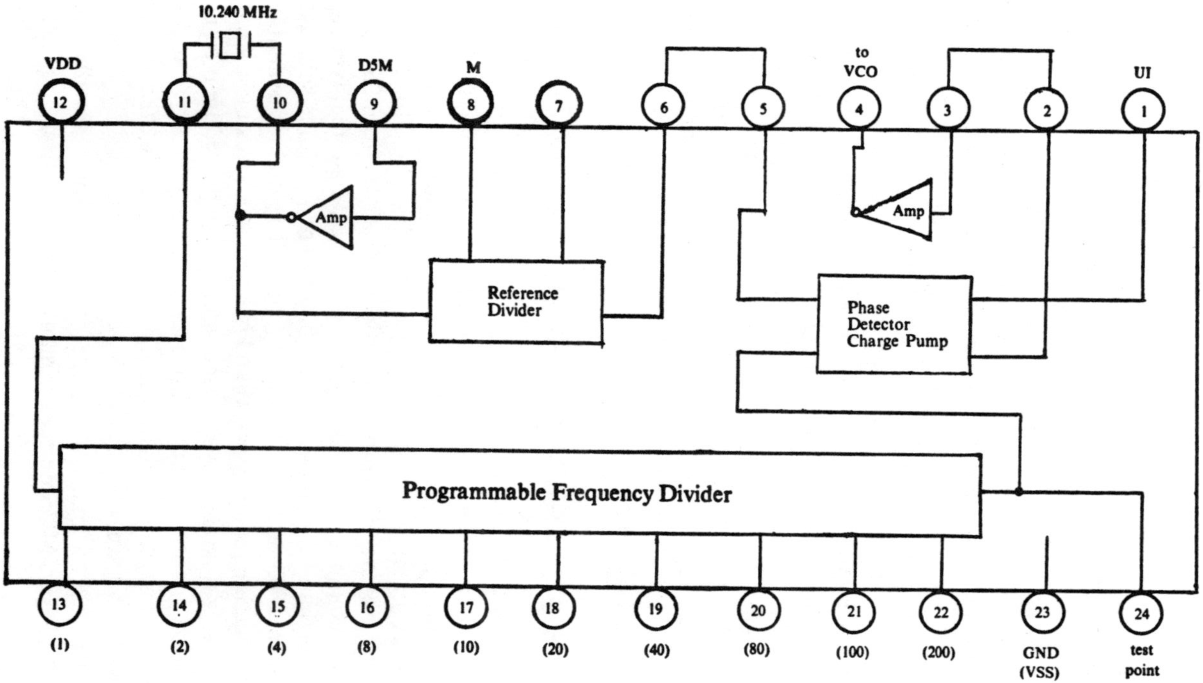

Fig. 7-2. A breakdown of the internal structure of the muPD858C.

oscillator with an external crystal. Pin 10 is the output of this amplifier. Figure 7-2 shows that the reference frequency crystal connects between the input and the output of the amplifier. Pin 11 is the input to the programmable divider. Pin 12 is the supply voltage to the IC. This is usually labeled VDD on CMOS devices.

Pins 13 through 22 are the BCD programming input lines to the programmable divider. The weight of each program line is shown below the pin numbers. The divisor N of the programmable divider is determined by the binary level on the various program lines. The maximum divisor N possible is 465, the sum of the various weights of the program lines. Each program line would have to be at the binary 1 level to program an N divisor of 465 into the programmable divider. To determine the specific binary code needed for a specific divisor N, proceed as follows. First, find the highest available weight of any program line that is equal to or less than the desired N value. Place binary 1 on that programming pin and subtract that weight from the desired N value. Next, find the highest available weight that is equal to or less than the remainder. Place binary 1 on that programming pin and repeat the above procedure until there is no remainder. Then place binary 0 on the unused pins. For example, suppose you want to program an N value of 325 into the programmable divider. The highest available weight that is equal to or less than 325 is 200 (pin 22). Mark binary 1 at pin 22 and subtract 200 from 325. This leaves a remainder of 125. The highest available weight which is equal to or less than 125 is 100 (pin 21). Mark binary 1 at pin 21 and subtract 100 from 125. This leaves a remainder of 25. The highest available weight which is equal to or less than 25 is 20 (pin 18). Mark binary 1 at pin 18 and the subtract 20 from 25. This leaves a remainder of 5. The highest available weight that is equal to or less than 5 is 4 (pin 15). Mark binary 1 at pin 15 and subtract 4 from 5. This leaves a remainder of 1. The highest available weight that is equal to or less than 1 *is* 1, (pin 13). Mark binary 1 at pin 13. This time there is no remainder, so the procedure is complete, except for one thing. Go back and mark binary 0 at the program pins whose weight was not used. What you then have is the BCD program code for an N value of 325.

Pin 23 is common or ground for the IC. This is usually labeled VSS on CMOS devices. Finally, pin 24 is a test point that connects to the output of the programmable divider. This can be a very handy test point when troubleshooting this PLL device. For example, you could determine whether or not the programmable divider is dividing properly by checking at pin 24 with a frequency counter.

PRACTICAL APPLICATION

Figure 7-3 shows a block diagram of a practical application of the muPD858C PLL IC. This same PLL circuit is used in the Cobra 89XLR CB base station.

The PLL IC pin connections show that the reference frequency select pin (pin 7) is connected to +4.7 volts. This sets the reference frequency at 10 kHz.

Let's next check the programming pins (13 through 22) to see how they are set up. Pins 13 through 19 connect to the channel selector switch and thus are at binary 1 or binary 0 level, depending upon the channel selector switch position. The three remaining program pins (20, 21, and 22) are not connected to the channel selector switch. Instead, pins 20 and 22 are connected directly to ground. This keeps these pins at the binary 0 level at all times. Pin 21 is connected directly to +4.7 volts, so it is kept at the binary 1 level at all times. Figure 7-2 shows the weights of all the program pins. Because pin 21 is always kept at the binary 1 level, its weight of 100 will always be used in determining the divisor N. The weights of pins 20 and 22 will never be used because they are kept at binary 0 (ground) at all times. Thus, the divisor N will be 100 plus whatever is programmed into the other program lines by the channel selector. Table 7-1 lists the N divisors and the programming code for each channel, 1 through 40. The pin weight is shown in parenthesis below the pin number. To determine what N number is programmed into the programmable divider, simply add the weights of the various program pins where the binary 1 appears. For example, the binary 1 level for channel 1 appears at pins 13, 16, and 21. The weights of these pins are 1, 8, and 100 respectively. The sum of these weights if 109, the N number for that channel.

As seen from Table 7-2, the N number ranges from 109 to 153 for CB channels 1 through 40. Table 7-2 shows the VCO frequency for each of the 40 CB channels. Also shown on the chart are the offset frequencies (the input to the programmable divider), the offset oscillator frequency (which is the same on all channels), and the N number for each channel. Notice that the offset frequency for each channel is equal to the VCO frequency for that channel minus the offset oscillator frequency (36.570 MHz).

As shown in Table 7-2, the VCO operates from 37.660 MHz to 38.100 MHz for channels 1 through 40. The VCO frequency is

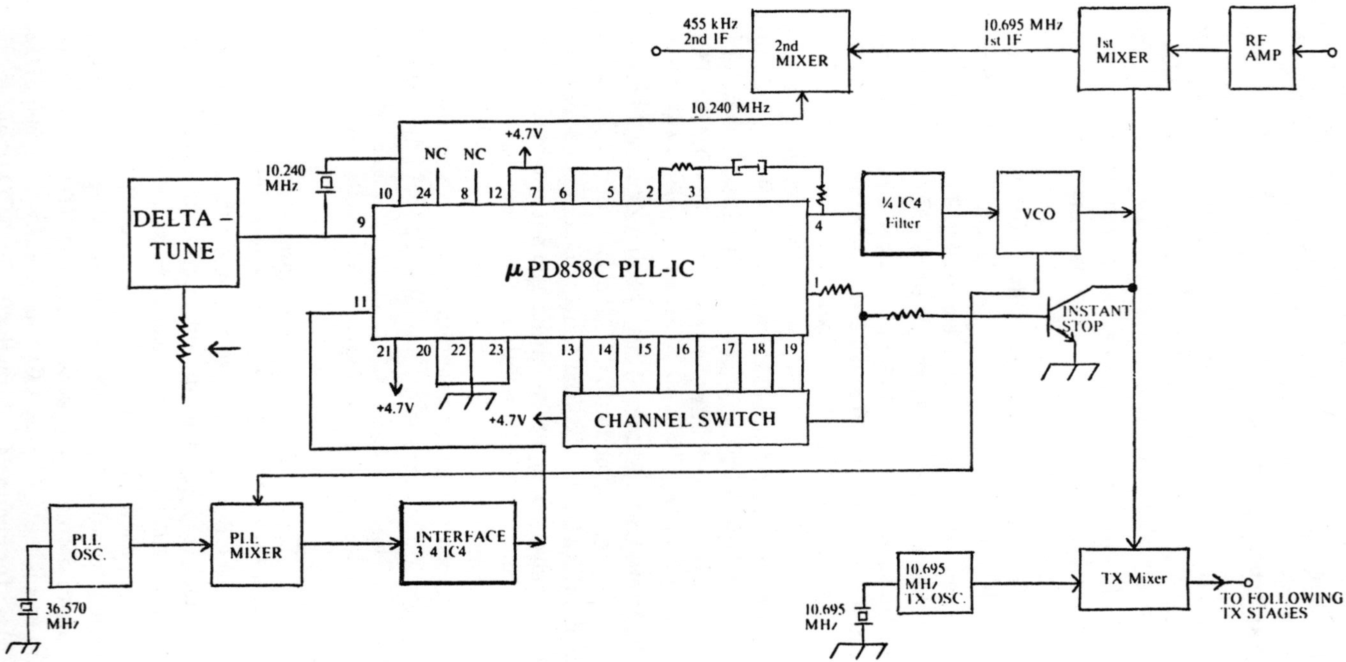

Fig. 7-3. Block diagram of a typical application of the μPD858C.

Table 7-1. N Numbers and Associated Program Codes for the PLL System of Fig. 7-3.

Channel No.	Divide No.	Programming code to programmable divider									
		Pin----------13	14	15	16	17	18	19	20	21	22
		Weight-------1	2	4	8	10	20	40	80	100	200
1	109	1	0	0	1	0	0	0	0	1	0
2	110	0	0	0	0	1	0	0	0	1	0
3	111	1	0	0	0	1	0	0	0	1	0
4	113	1	1	0	0	1	0	0	0	1	0
5	114	0	0	1	0	1	0	0	0	1	0
6	115	1	0	1	0	1	0	0	0	1	0
7	116	0	1	1	0	1	0	0	0	1	0
8	118	0	0	0	1	1	0	0	0	1	0
9	119	1	0	0	1	1	0	0	0	1	0
10	120	0	0	0	0	0	1	0	0	1	0
11	121	1	0	0	0	0	1	0	0	1	0
12	123	1	1	0	0	0	1	0	0	1	0
13	124	0	0	1	0	0	1	0	0	1	0
14	125	1	0	1	0	0	1	0	0	1	0
15	126	0	1	1	0	0	1	0	0	1	0
16	128	0	0	0	1	0	1	0	0	1	0
17	129	1	0	0	1	0	1	0	0	1	0
18	130	0	0	0	0	1	1	0	0	1	0
19	131	1	0	0	0	1	1	0	0	1	0
20	133	1	1	0	0	1	1	0	0	1	0
21	134	0	0	1	0	1	1	0	0	1	0
22	135	1	0	1	0	1	1	0	0	1	0
23	138	0	0	0	1	1	1	0	0	1	0
24	136	0	1	1	0	1	1	0	0	1	0
25	137	1	1	1	0	1	1	0	0	1	0
26	139	1	0	0	1	1	1	0	0	1	0
27	140	0	0	0	0	0	0	1	0	1	0
28	141	1	0	0	0	0	0	1	0	1	0
29	142	0	1	0	0	0	0	1	0	1	0
30	143	1	1	0	0	0	0	1	0	1	0
31	144	0	0	1	0	0	0	1	0	1	0
32	145	1	0	1	0	0	0	1	0	1	0
33	146	0	1	1	0	0	0	1	0	1	0
34	147	1	1	1	0	0	0	1	0	1	0
35	148	0	0	0	1	0	0	1	0	1	0
36	149	1	0	0	1	0	0	1	0	1	0
37	150	0	0	0	0	1	0	1	0	1	0
38	151	1	0	0	0	1	0	1	0	1	0
39	152	0	1	0	0	1	0	1	0	1	0
40	153	1	1	0	0	1	0	1	0	1	0

Table 7-2. Various Specifications of the PLL System of Fig. 7-3.

Channel No	Divide Ratio (N)	F-Offset Freq (MHz)	F-Offset Osc. Freq (MHz)	Fvco Freq (MHz)
1	109	1.09	36.570	37.660
2	110	1.10	"	37.670
3	111	1.11	"	37.680
4	113	1.13	"	37.700
5	114	1.14	"	37.710
6	115	1.15	"	37.720
7	116	1.16	"	37.730
8	118	1.18	"	37.750
9	119	1.19	"	37.760
10	120	1.20	"	37.770
11	121	1.21	"	37.780
12	123	1.23	"	37.800
13	124	1.24	"	37.810
14	125	1.25	"	37.820
15	126	1.26	"	37.830
16	128	1.28	"	37.850
17	129	1.29	"	37.860
18	130	1.30	"	37.870
19	131	1.31	"	37.880
20	133	1.33	"	37.900
21	134	1.34	"	37.910
22	135	1.35	"	37.920
23	138	1.38	"	37.950
24	136	1.36	"	37.930
25	137	1.37	"	37.940
26	139	1.39	"	37.960
27	140	1.40	"	37.970
28	141	1.41	"	37.980
29	142	1.42	"	37.990
30	143	1.43	"	38.000
31	144	1.44	"	38.010
32	145	1.45	"	38.020
33	146	1.46	"	38.030
34	147	1.47	"	38.040
35	148	1.48	"	38.050
36	149	1.49	"	38.060
37	150	1.50	"	38.070
38	151	1.51	"	38.080
39	152	1.52	"	38.090
40	153	1.53	"	38.100

mixed with the 36.570-MHz PLL oscillator (offset oscillator) signal in the PLL mixer (offset mixer). The PLL mixer is a dual gate MOS N-channel field-effect transistor (FET). The VCO signal is fed to one gate, while the PLL oscillator signal is fed to the other gate. The two signals are hererodyned inside the FET transistor. Since there is a low-pass filter connected in the drain circuit of the FET, only the difference frequency (1.09 MHz through 1.53 MHz) is allowed to pass. The other frequencies are sharply attenuated by the filter.

From the low-pass filter, the offset frequency is fed to IC4. This is a CMOS quad 2-input NAND gate. Figure 7-4 shows the internal structure of IC4. The diagram also shows the way the IC is connected in the PLL circuit, so that its action can be more easily analyzed. Notice that both inputs of each NAND gate are connected. This simply makes the NAND gate an *inverter.* Three of the NAND gates)2, 3, and 4) are connected in cascade between the output of the PLL mixer and the input of the programmable divider. The other gate (1) is used as a filter between pin 4 of IC3 and the VCO control line (pin 5 of IC5).

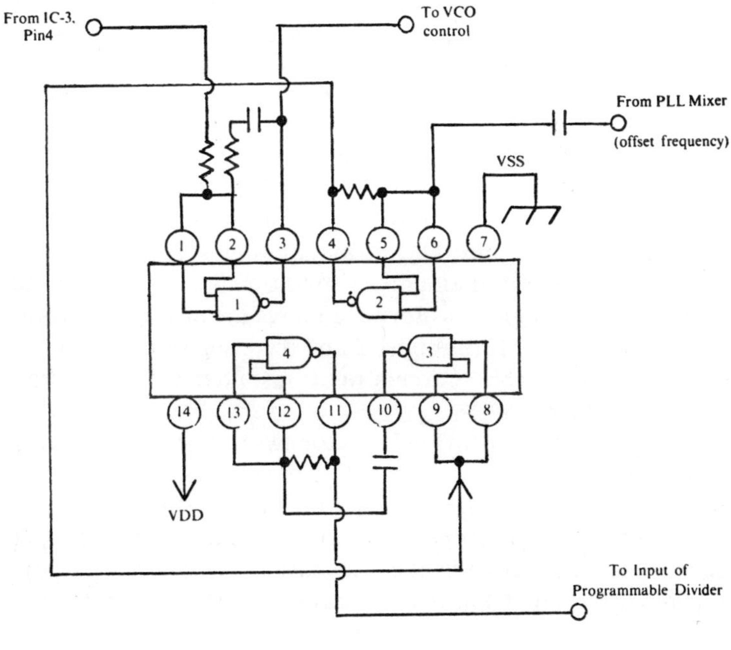

Fig. 7-4. A breakdown of the internal structure of the MB84011U CMOS quad NAND gate IC.

The programmable divider divides its input frequency by N. If the output of the programmable divider is not exactly 10 kHz, a correction signal (after being filtered) is applied to the control line of the VCO (pin 5 of IC5). This correction signal moves the VCO frequency until the output of the programmable divider is exactly 10 kHz. When the output of the programmable divider reaches 10 kHz, the control voltage stabilizes the VCO frequency.

The VCO used in this PLL circuit is an IC, a UH1C004. Figure 7-5A shows the diagram of the IC. Figure 7-5B shows a sketch of the IC with the pin identified. Pin 1 of the UH1C004 IC is the buffered output. Pin 2 is ground. Pin 3 is no connection. Pin 4 connects to an external coil for setting the VCO range. Pin 5 is the control line of the VCO oscillator. Pins 6 and 7 are not connected. Pin 8 connects to the supply voltage. Pin 9 is normally connected to ground. Pin 10 is the output from the VCO collector.

In the receive mode, the VCO signal is fed to the first mixer, where it is heterodyned with the incoming receive signal to produce a 10.695-MHz first intermediate frequency. The 10.695-MHz first i-f is then fed to the second mixer, where it is heterodyned with a 10.240-MHz signal from the reference oscillator. The difference of these two frequencies (455 kHz) is the second i-f.

In the transmit mode, the VCO frequency is fed to the transmit mixer. Where, it is heterodyned with a 10.695-MHz signal from the transmit oscillator. The difference of these two frequencies is the transmit frequency.

Notice in the block diagram of Fig. 7-3 that a transistor (labeled *instant stop*) connects to the output of the VCO. Two inputs feed the base of the transistor: one from the lock detector output (pin 1) of the muPD858C and another from the channel switch. In the event that an unlock condition occurs, pin 1 of the muPD858C goes high. This forward biases the instant stop transistor. The transistor turns on hard and therefore shunts the VCO signal to ground. This prevents the transmitter from radiating improper frequencies.

It is possible for a channel selector switch to apply improper programming to the programmable divider, either because of a defective channel switch or by the operator setting the switch between channels. Improper programming causes an improper frequency to be radiated by the transmitter. To prevent this from happening, the line from the channel switch to the instant stop transistor goes HIGH when the channel selector malfunctions or is improperly set. This turns on the instant stop transistor and thus kills the VCO output.

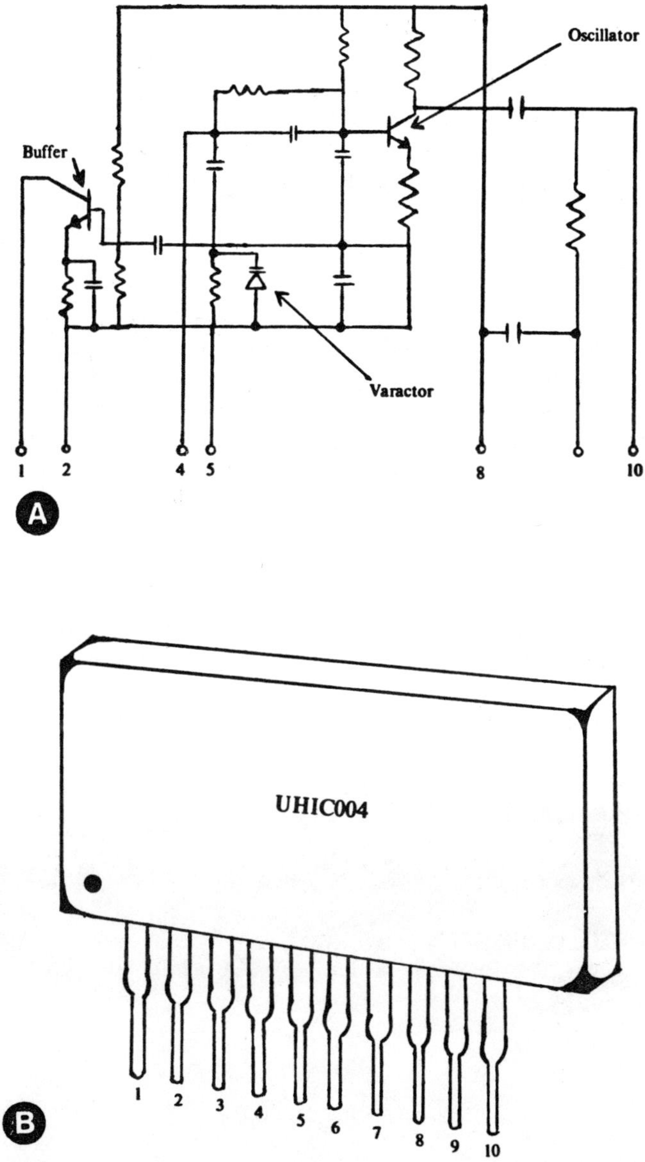

Fig. 7-5A & B. Complete diagram of the μHIC004 VCO/VCO buffer IC is shown at A, and the μHIC004 IC pin identification is shown at B (courtesy of Cobra Communications).

Here's how the PLL circuit works for channel 1. On channel 1, the programmable divider is programmed to divide its input by 109. Therefore, for a 10-kHz output, the input must be 1.09 MHz. In order for the programmable divider input frequency to be 1.09 MHz, the VCO frequency must be 36.570 MHz + 1.09 MHz, or 37.660 MHz. If the VCO frequency is not 37.660 MHz, the phase detector within the muPD858C will produce an error signal that is used to steer the VCO to the proper frequency.

PRESIDENT MODEL MADISON

The President Model Madison CB radio is pictured in Fig. 7-6. This is a high-quality, state-of-the-art CB transceiver that is capable of operating in upper sideband (USB), lower sideband (LSB), or AM modes. The PLL synthesizer circuit is designed around the muPD858C PLL IC. It uses only six crystals to generate all the frequencies required for receiving and transmitting in any of the three color modes: USB, LSB or AM.

The schematic of the PLL circuit is shown in Fig. 7-7. The remainder of the synthesizer and related circuits are shown in Fig. 7-8. First, the operation and function of the various stages will be explained and then the complete circuit operation with mathematical analysis will be presented.

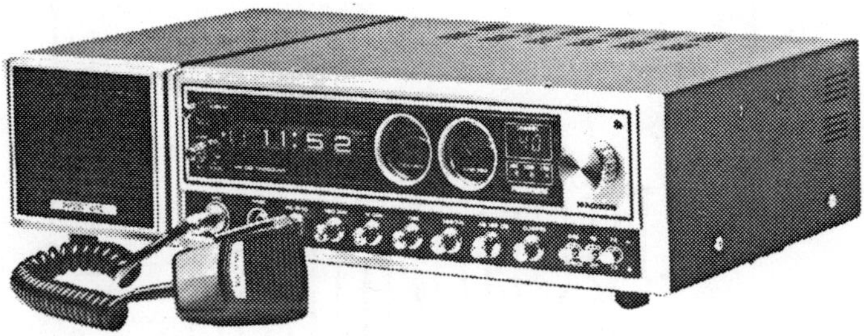

Fig. 7-6. The President Model Madison (courtesy of President Electronics/ American Radio).

μPD858C IC PLL Stage

First, checking the reference frequency select pin (pin 7) you can see that it is connected to +4.7 volts. This means that a reference frequency of 10 kHz has been selected. Next check the programming lines (pins 13 through 22) to see how they are set up. Actually, eight of the program lines connect to the channel switch. These are pins 13 through 18 and pins 20 and 21. The other two program lines (pins 19 and 22) are connected directly to ground, thus keeping these pins at binary 0 at all times. This means that the weights of pins 19 and 22 (40 and 200, respectively) will never be used in determining the divisor N of the programmable divider. In this particular PLL system, the N divisor ranges from 91 for channel 1 to 135 for channel 40. The divisor N and the programming code for each channel is listed in Table 7-3. Figure 7-7 shows that each of the program lines which connect to the channel selector switch each have a pulldown resistor connected between the program line and ground. This is actually eight resistors in one package. One of the pins is common. The other eight pins connect to the individual resistors. This resistor package is labeled RR1 on the schematic. These pulldown resistors simplify the design of the channel selector switch. The channel selector simply connects the proper program lines to the binary 1 supply point (+4.7 volts). The program lines that are left open by the channel selector are placed at ground potential (binary 0) by the pulldown resistors. Without the pulldown resistors, the channel selector would have to be designed to ground the program lines that should be at the binary 0 level.

Pin 2 (output of phase detector/charge pump) is connected to pin 3 (the input to the on-chip active filter) through a 15K resistor. Pin 4 is the output of the active filter. Notice the RC network between the input and the output of the active filter. This RC network, along with the on-chip amplifier, forms a low-pass filter that cleans up the control signal before it is applied to the VCO control line.

VCO Stage

The VCO stage uses an IC that contains the VCO oscillator with an on-chip varactor diode. The IC also contains a buffer stage. The IC is a UHIC005A. It is very similar to the UHIC004A IC that was discussed previously in this chapter. Notice that coil L17 connects to pin 4. This coil can be considered as a course frequency adjustment for the VCO. This coil is tuned to set the frequency of the VCO "in the ballpark." Fine tuning of the VCO is done through the control voltage to the VCO control line. The VCO control line is brought out to pin 5 of IC5 (see Fig. 7-7). The VCO operates in the 35-MHz range. Tables 7-4 and. 7-5 list the

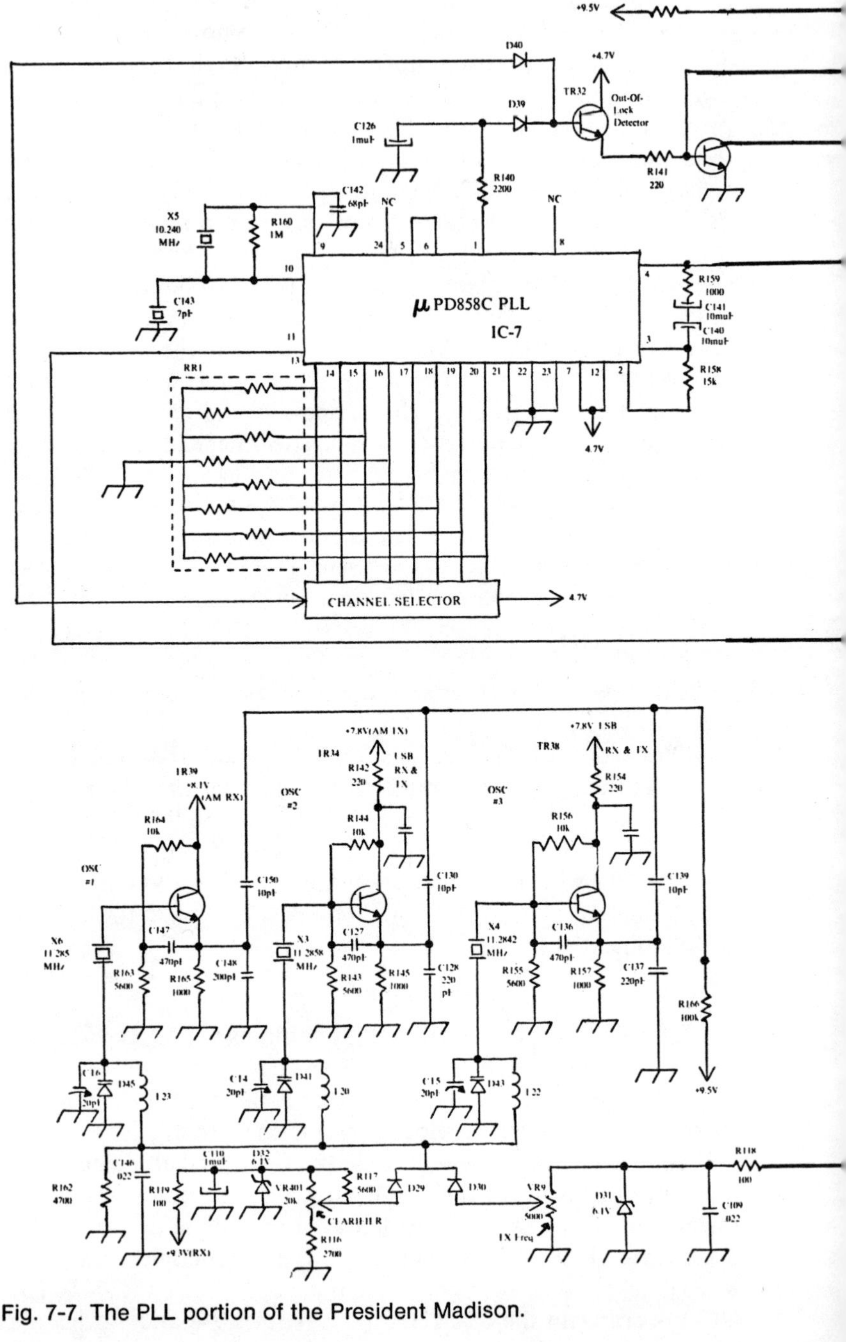

Fig. 7-7. The PLL portion of the President Madison.

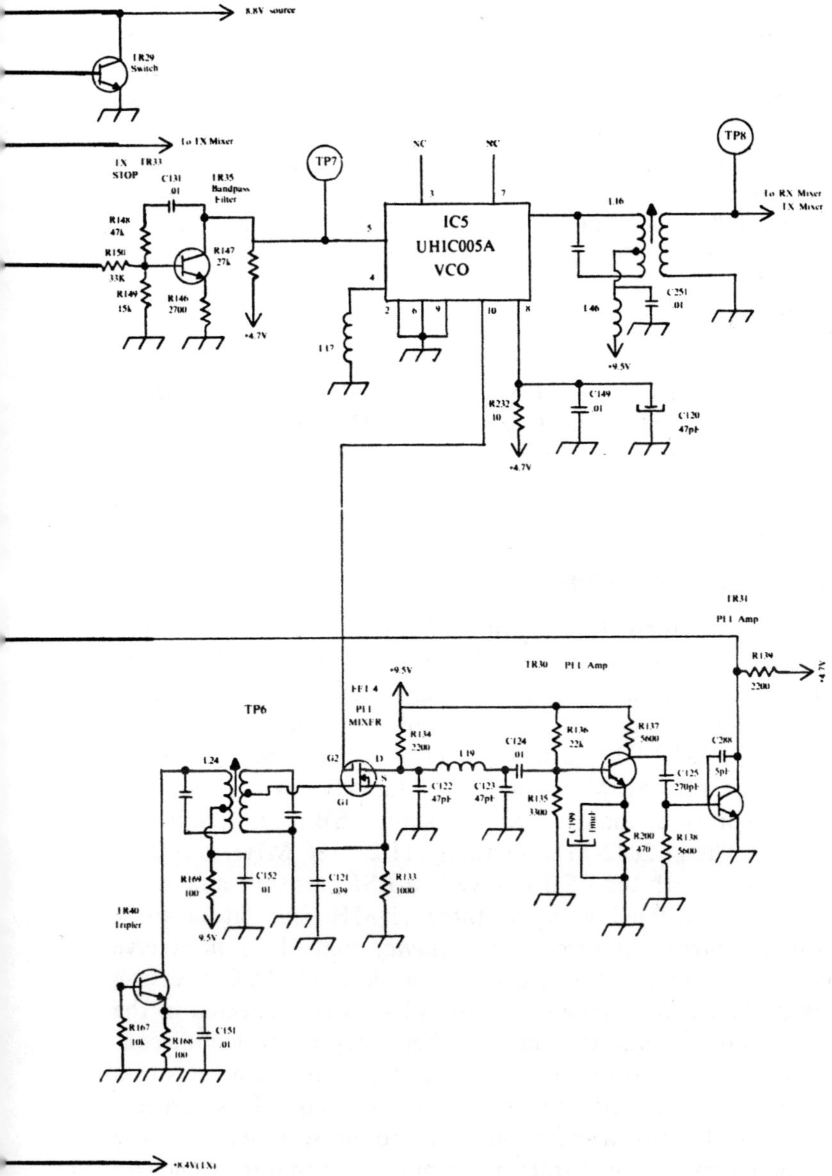
8.8V source
TR29
Switch
To TX Mixer
TX STOP
TR33
C131
.01
TR35
Bandpass
Filter
R148
47k
R150
33K
R149
15k
R146
2700
R147
27k
+4.7V
TP7
NC
NC
IC5
UHIC005A
VCO
L17
TP8
To RX Mixer
TX Mixer
L16
C251
.01
L46
+9.5V
R232
10
C149
.01
C120
47pF
+4.7V
TR31
PLL Amp
R139
2200
+4.7V
TR30
PLL Amp
+9.5V
FET 4
PLL
MIXER
TP6
L24
R134
2200
L19
C124
.01
R136
22k
R137
5600
C288
5pF
C122
47pF
C123
47pF
R135
3300
C125
270pF
C199
1mmf
R200
470
R138
5600
R169
100
C152
.01
C121
.039
R133
1000
9.5V
TR40
Tripler
R167
10k
R168
100
C151
.01
+8.4V(TX)

VCO frequency for each channel in the various modes of operation.

One VCO signal is fed from pin 10 of IC5 to one gate of the PLL mixer FET. Here, the VCO signal is mixed with a crystal-controlled frequency to produce an offset frequency that is equal to the VCO frequency minus the crystal-controlled oscillator frequency. Another VCO signal, which is amplified, is fed from pin 1 of IC5 to transformer L16. From the secondary of L16, the VCO signal is fed to the first mixer and to the transmit mixer (See Figs. 7-7 and 7-8).

Bandpass Filter

The bandpass filter (TR35) is connected between the output of the active filter (part of IC7) and the VCO control line (pin 5 of IC5). The function of this stage is to provide additional loop filtering. The RC filter network is connected between the base and collector of TR35.

Offset Crystal Oscillators

There are three different offset frequency oscillators used in this set. The output of each oscillator feeds the input of a tripler (TR40). The oscillator frequencies are thus multiplied by 3 to yield a 33-MHz signal at the output of the tripler.

The 11.285-MHz oscillator functions only in the AM receive mode. The 11.285B-MHz (actually, 11.285833-MHz) oscillator functions in three modes: USB transmit, USB receive, and AM transmit. The 11.2842-MHz (actually, 11.284166-MHz) oscillator functions in both the LSB receive and LSB transmit modes.

Also associated with the three 11-MHz oscillators are a *clarifier* control and a *transmit frequency* control. In the receive mode, diode D29 is forward biased while diode D30 is reverse biased. Thus, the clarifier control (VR401) is connected in the receive mode, while the transmit frequency control is disconnected. Notice that the junction of D29 and D30 connects to a varactor in each oscillator through a choke coil. This varactor is connected between the crystal and ground so that changes in the capacitance of the varactor can pull the crystal frequency up or down slightly. Notice that the top end of the clarifier control is connected to a supply voltage that is regulated by the 6.1-volt zener, D32. As the clarifier control is varied, the reverse bias to the varactor diodes also varies. This change pulls the oscillator

Table 7-3. N Numbers and Associated Program Codes for the President Model Madison PLL Circuit.

Channel No.	Divisor N	Programming code to the programmable divider									
		Pin------13	14	15	16	17	18	19	20	21	22
		Weight---1	2	4	8	10	20	40	80	100	200
1	91	1	0	0	0	1	0	0	1	0	0
2	92	0	1	0	0	1	0	0	1	0	0
3	93	1	1	0	0	1	0	0	1	0	0
4	95	1	0	1	0	1	0	0	1	0	0
5	96	0	1	1	0	1	0	0	1	0	0
6	97	1	1	1	0	1	0	0	1	0	0
7	98	0	0	0	1	1	0	0	1	0	0
8	100	0	0	0	0	0	0	0	0	1	0
9	101	1	0	0	0	0	0	0	0	1	0
10	102	0	1	0	0	0	0	0	0	1	0
11	103	1	1	0	0	0	0	0	0	1	0
12	105	1	0	1	0	0	0	0	0	1	0
13	106	0	1	1	0	0	0	0	0	1	0
14	107	1	1	1	0	0	0	0	0	1	0
15	108	0	0	0	1	0	0	0	0	1	0
16	110	0	0	0	0	1	0	0	0	1	0
17	111	1	0	0	0	1	0	0	0	1	0
18	112	0	1	0	0	1	0	0	0	1	0
19	113	1	1	0	0	1	0	0	0	1	0
20	115	1	0	1	0	1	0	0	0	1	0
21	116	0	1	1	0	1	0	0	0	1	0
22	117	1	1	1	0	1	0	0	0	1	0
23	120	0	0	0	0	0	1	0	0	1	0
24	118	0	0	0	1	1	0	0	0	1	0
25	119	1	0	0	1	1	0	0	0	1	0
26	121	1	0	0	0	0	1	0	0	1	0
27	122	0	1	0	0	0	1	0	0	1	0
28	123	1	1	0	0	0	1	0	0	1	0
29	124	0	0	1	0	0	1	0	0	1	0
30	125	1	0	1	0	0	1	0	0	1	0
31	126	0	1	1	0	0	1	0	0	1	0
32	127	1	1	1	0	0	1	0	0	1	0
33	128	0	0	0	1	0	1	0	0	1	0
34	129	1	0	0	1	0	1	0	0	1	0
35	130	0	0	0	0	1	1	0	0	1	0
36	131	1	0	0	0	1	1	0	0	1	0
37	132	0	1	0	0	1	1	0	0	1	0
38	133	1	1	0	0	1	1	0	0	1	0
39	134	0	0	1	0	1	1	0	0	1	0
40	135	1	0	1	0	1	1	0	0	1	0

Table 7-4. President Madison PLL Circuit Transmit Code Specifications.

Channel	Divide Ratio (N)	VCO Frequencies in MHz USB & AM	VCO Frequencies in MHz LSB	Carrier Oscillator Frequencies MHz AM & USB	Carrier Oscillator Frequencies MHz LSB
1	91	34.7675	34.7625	7.8025	7.7975
2	92	34.7775	34.7725	"	"
3	93	34.7875	34.7825	"	"
4	95	34.8075	34.8025	"	"
5	96	34.8175	34.8125	"	"
6	97	34.8275	34.8225	"	"
7	98	34.8375	34.8325	"	"
8	100	34.8575	34.8525	"	"
9	101	34.8675	34.8625	"	"
10	102	34.8775	34.8725	"	"
11	103	34.8875	34.8825	"	"
12	105	34.9075	34.9025	"	"
13	106	34.9175	34.9125	"	"
14	107	34.9275	34.9225	"	"
15	108	34.9375	34.9325	"	"
16	110	34.9575	34.9525	"	"
17	111	34.9675	34.9625	"	"
18	112	34.9775	34.9725	"	"
19	113	34.9875	34.9825	"	"
20	115	35.0075	35.0025	"	"
21	116	35.0175	35.0125	"	"
22	117	35.0275	35.0225	"	"
23	120	35.0575	35.0525	"	"
24	118	35.0375	35.0325	"	"
25	119	35.0475	35.0425	"	"
26	121	35.0675	35.0625	"	"
27	122	35.0775	35.0725	"	"
28	123	35.0875	35.0825	"	"
29	124	35.0975	35.0925	"	"
30	125	35.1075	35.1025	"	"
31	126	35.1175	35.1125	"	"
32	127	35.1275	35.1225	"	"
33	128	35.1375	35.1325	"	"
34	129	35.1475	35.1425	"	"
35	130	35.1575	35.1525	"	"
36	131	35.1675	35.1625	"	"
37	132	35.1775	35.1725	"	"
38	133	35.1875	35.1825	"	"
39	134	25.1975	35.1925	"	"
40	135	35.2075	35.2025	"	"

Table 7-5. President Madison PLL Circuit Receive Mode Code Specifications.

CH.	Divide Ratio (N)	VCO Frequencies in MHz AM	USB	LSB	IF Frequency (MHz)
1	91	34.765	34.7675	34.7625	7.8
2	92	34.775	34.7775	34.7725	″
3	93	34.785	34.7875	34.7825	″
4	95	34.805	34.8075	34.8025	″
5	96	34.815	34.8175	34.8125	″
6	97	34.825	34.8275	34.8225	″
7	98	34.835	34.8375	34.8325	″
8	100	34.855	34.8575	34.8525	″
9	101	34.865	34.8675	34.8625	″
10	102	34.875	34.8775	34.8725	″
11	103	34.885	34.8875	34.8825	″
12	105	34.905	34.9075	34.9025	″
13	106	34.915	34.9175	34.9125	″
14	107	34.925	34.9275	34.9225	″
15	108	34.935	34.9375	34.9325	″
16	110	34.955	34.9575	34.9525	″
17	111	34.965	34.9675	34.9625	″
18	112	34.975	34.9775	34.9725	″
19	113	34.985	34.9875	34.9825	″
20	115	35.005	35.0075	35.0025	″
21	116	35.015	35.0175	35.0125	
22	117	35.025	35.0275	35.0225	″
23	120	35.055	35.0575	35.0525	″
24	118	35.035	35.0375	35.0325	″
25	119	35.045	35.0475	35.0425	″
26	121	35.065	35.0675	35.0625	″
27	122	35.075	35.0775	35.0725	″
28	123	35.085	35.0875	35.0825	″
29	124	35.095	35.0975	35.0925	″
30	125	35.105	35.1075	35.1025	″
31	126	35.115	35.1175	35.1125	″
32	127	35.125	35.1275	35.1225	″
33	128	35.135	35.1375	35.1325	″
34	129	35.145	35.1475	35.1425	″
35	130	35.155	35.1575	35.1525	″
36	131	35.165	35.1675	35.1625	″
37	132	35.175	35.1775	35.1725	″
38	133	35.185	35.1875	35.1825	″
39	134	35.195	35.1975	35.1925	″
40	135	35.205	35.2075	35.2025	″

frequency. This clarifier voltage is applied to all three varactor diodes (D45, D41, and D43) simultaneously, but only one of the oscillators is functional at a time. Which one depends upon the mode of operation. In the AM mode, the clarifier is more of a luxury. In the SSB mode, however, the clarifier is a necessity. Otherwise, slightly off-frequency sideband signals might be completely unreadable.

In the transmit mode, diode D29 becomes reverse biased (isolating the clarifier control), and D30 becomes forward biased (connecting the transmit frequency control (VR9) to the varactor control line). The voltage to the transmit frequency control is 6.1 volts, regulated by zener D31. In the transmit mode, therefore, the control voltage to the varactor is placed at a fixed level that returns the oscillator to its center frequency, regardless of the position of the clarifier control. The transmit frequency control is an internal adjustment, not accessible to the operator.

As mentioned, the 11-MHz oscillator frequency is tripled by TR40. The primary of transformer L24 (in the collector circuit of TR40) is tuned to the third harmonic of the 11-MHz oscillators (approximately 33.8 MHz). This 33-MHz signal is then fed from the secondary of L24 to one input of the PLL or offset mixer.

PLL Mixer

The PLL, or offset, mixer is a dual-gate MOSFET transistor. The offset oscillator frequency feeds one gate of the FET (G1), and the VCO signal feeds the other gate (G2). These two frequencies are mixed. The difference of these two frequencies appears at the output (drain) of the transistor. The difference frequency is equal to the VCO frequency minus the offset (PLL) oscillator frequency. The other frequencies are filtered by the low-pass filter that is composed of C122, L19, and C123.

PLL Amplifier

The PLL amplifier consists of two transistor stages, TR30 and TR31. This stage amplifies the offset frequency to a sufficient level to feed the input to the programmable divider.

Out-of-Lock Detector/Transmit Stop Circuit

The out-of-lock detector transistor (TR32) has two inputs, one from the lock detector output at pin 1 of IC7 and the other from the channel selector switch. If the channel selector switch is improperly set between channels, this would cause improper pro-

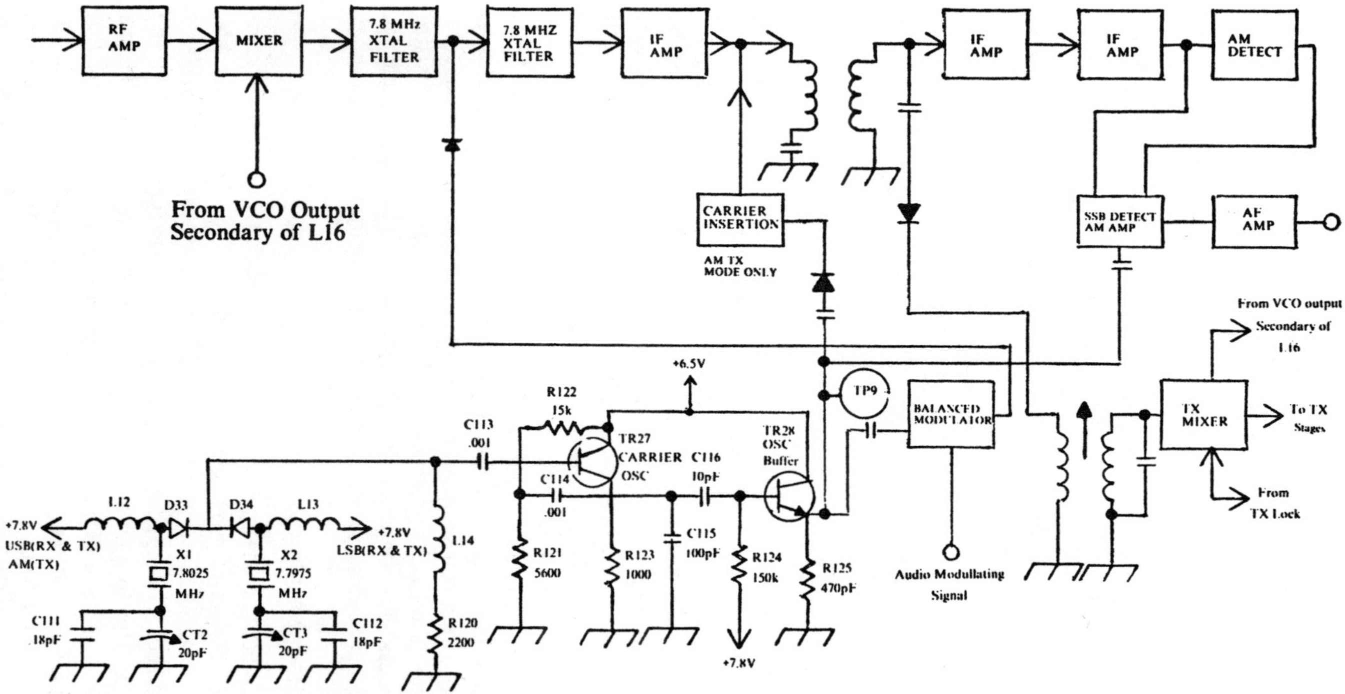

Fig. 7-8. Partial schematic/block diagram of President Madison, showing carrier oscillator and other stages pertinent to the PLL synthesizer discussion.

gramming to the μPD858C. This might in turn result in the generation of an improper frequency that might not be a legal CB frequency. To prevent the set from transmitting an improper frequency, when the selector is improperly positioned, a HIGH level voltage will be placed on the line going from the channel selector switch to diode D40 at the input of TR32. This high voltage level will turn on TR32, which in turn causes both TR29 and TR33 to turn on. When TR33 turns on, it kills the supply voltage to the drain of the FET transmit mixer. When TR29 turns on, it kills the supply voltage to the carrier insertion transistor. Thus, the transmitter is disabled.

The output from the lock detector of IC7, pin 1, causes the same sequence of events. During a condition of unlock, a HIGH level voltage will appear at pin 1 of IC7. This will turn on TR32, causing TR33 and TR29 to turn on and killing the transmitter.

LED Channel Display Circuit

The LED channel display has two seven-segment LEDs on one chip, one for the tens digit and the other for the units digit. The rear view drawing of the display is shown in Fig. 7-9. Figure 7-10 shows the various segment connections. This display is of the common-anode type. Notice that the common terminals (12 and 13) connect to the positive supply voltage through diode D74 and transistor TR403. Each segment of the LED is supplied approximately 2 volts at 12 milliamperes.

Notice that two "packages" of resistors are used with the LED display. Each package contains seven resistors, one for each segment. The purpose of these resistors is to limit the current flow through the LED segment to a safe level. One end of each resistor connects back to the channel selector switch. The channel selector switch provides a ground for the proper resistors for each channel.

AM Receive Mode

When the channel selector is set for channel 40, the programmable divider is programmed to divide by 135. Also, in the AM receive mode, offset oscillator 1 (which operates at 11.285 MHz) is energized. This oscillator signal is fed to the base of TR40, the tripler. The frequency at the output of the tripler is 3 × 11.285 MHz, or 33.855 MHz. This 33.855-MHz signal is fed to gate 1 of the FET PLL mixer stage. Here, it is mixed with the signal from the VCO (IC5) to produce a difference frequency at the output (drain) of the FET PLL mixer (FET 4). If you represent the output

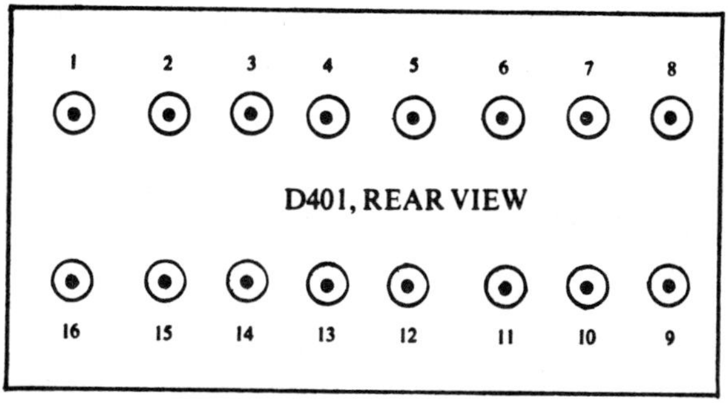

Fig. 7-9. A rear view of the LED channel display that shows the pin identification numbers.

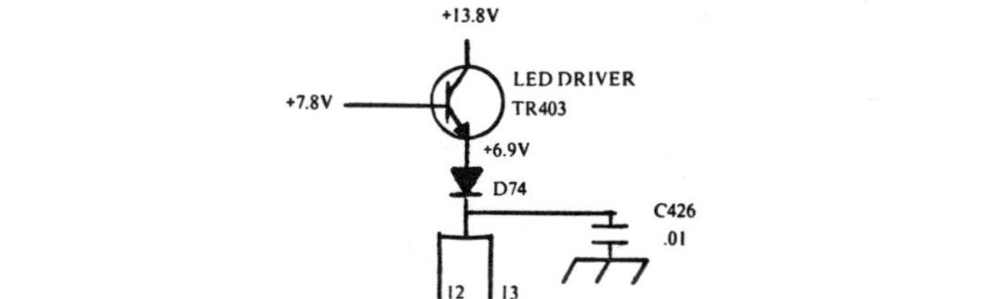

Fig. 7-10. The LED display and associated circuitry.

of the tripler (TR40) as $f\times3$ and the VCO frequency as fVCO, then the formula for the VCO frequency would be $fVCO = f\times3 + N \times N \times 10^{-2}$. We have already established that in the AM receive mode, the output of the tripler (TR40) is 33.855 MHz and the divisor N is 135. Substituting these values into the formula, you can determine the VCO frequency: fVCO (MHz) = 33.855 + 135 × 10^{-2} = 33.855 + 1.35, or 35.205 MHz. If the VCO frequency is 35.205 MHz, the input to the programmable divider will be 35.205 MHz-33.855 MHz, or 1.35 MHz. The output of the programmable divider will then be 1.35 MHz ÷135, or 10 kHz. Because this is equal to the standard reference frequency, the loop will be locked. If the VCO frequency were not exactly 35.205 MHz, the output of the programmable divider would not be 10 kHz. In this case, the phase detector would produce an error signal which after filtering would be used to correct the VCO frequency by changing the VCO control line voltage.

Now that you see how the AM receive mode VCO frequency is developed, let's take a look at how it is used. The 35.205-MHz VCO signal is fed to the receive mixer (shown in Fig. 7-8), where it is mixed with a 27.405-MHz channel 40 signal to produce a difference frequency of 35.205 MHz-27.405 MHz, or 7.8 MHz at the output of the mixer. This 7.8-MHz signal is the intermediate frequency. This 7.8-MHz i-f is then amplified and fed to the AM detector which recovers the audio signal.

AM Transmit Mode

In the AM transmit mode, oscillator 2 is energized, while oscillators 1 and 3 are not. Oscillator 2 operates at 11.285833 MHz. The output of the tripler, TR40, is then 3 × 285833 MHz, or 33.8575 MHz. The divisor N is still 135. From the formula, the VCO frequency is fVCO = 33.8575 MHz + 135 × 10^{-2}, or 35.2075 MHz. This 35.2075-MHz signal is fed to the transmit mixer (FET 7). In the AM transmit mode, the carrier oscillator (Fig. 7-8) operates at 7.8025 MHz. This signal is also fed to the transmit mixer (FET 7). The difference frequency of these two signals appears at the output of FET 7. This frequency is 35.2075 MHz - 7.8025 MHz, or 27.405 MHz. This is the transmit frequency for channel 40.

Upper Sideband Receive Mode

In the USB receive mode, oscillator 2 is energized. Because this is the same oscillator that was energized in the AM transmit mode, the VCO frequency will be the same as in the AM transmit mode, 35.2075 MHz. In the USB receive mode, this 35.2075-MHz signal is fed to the receive mixer, where it is heterodyned with the incoming receive signal.

Suppose that there is a channel 40 USB signal on the air being modulated by a 300 to 3000 Hz audio signal. This signal will then have a frequency range of 27.4053 MHz to 27.408 MHz and will be amplied and fed to the receive mixer. Here, it is mixed with the VCO signal to produce a difference frequency that ranges from 7.8022 MHz to 7.7995 MHz. Notice that the 27-MHz USB signal is converted in the mixer to a 7-MHz LSB signal. Anytime an SSB signal is mixed with a higher frequency signal to produce a difference frequency, the sidebands of the difference frequency will be inverted. This signal is passed by the 7.8-MHz crystal filters and then amplified and fed to the SSB product detector. Also, a 7.8025-MHz signal from the carrier oscillator is fed to the SSB product detector. The difference of these two frequencies appears at the output of the product detector. This frequency ranges from 300 to 3000 Hz. Thus, the original modulating frequencies are recovered. Table 7-6 shows how the channel 40 USB signal is processed in the receive mode.

Table 7-6. USB Receive Mode Frequency Development.

USB RECEIVE MODE FREQUENCY DEVELOPMENT

Conditions: Receiving channel 40 USB signal modulated by 300 Hz to 3000 Hz audio signal.

RECEIVE MIXER

VCO Freq. =	35.207500 MHz	35.207500 MHz
Rx Signal =	27.405300 MHz to	27.408000 MHz
IF Signal =	7.80220 MHz to	7.799500 MHz

SSB PRODUCT DETECTOR

Carrier Osc. Freq. =	7.802500 MHz	7.802500 MHz
IF Signal =	7.802200 MHz to	7.799500 MHz
Detected audio Freq =	300 Hz to	3000 Hz

Upper Sideband Transmit Mode

In the USB transmit mode, oscillator 2 is again energized. Because this is the same oscillator that was energized in the AM transmit and USB receive modes, the VCO frequency will be the same as in those modes, 35.2075 MHz. In the USB transmit mode, this 35.2075-MHz signal is fed to the transmit mixer (FET 7). Also, in the USB transmit mode, the carrier oscillator operates at 7.8025 MHz. This signal is buffered by TR28 and then fed to the balanced modulator (IC2). Also, the audio modulating signal is fed to the balanced modulator. With no audio input to the balanced modulator, there will be no output. If an audio signal is fed to the balanced modulator, however, the balanced modulator will produce a double sideband, suppressed-carrier output signal. The upper sideband signal will be 7.8025 MHz plus the audio modulating frequency. The lower sideband signal will be 7.8025 MHz plus the audio modulating frequency.

Say that the audio modulating frequency ranges from 300 to 3000 Hz. The upper sideband signal at the output of the balanced mixer will range from 7.8028 MHz to 7.8055 MHz. The lower sideband signal will range from 7.8022 MHz to 7.7995 MHz. The double sideband signal is fed to the 7.8-MHz filter, but only the lower sideband will be passed by the filter. The upper sideband signal will be eliminated. Thus, the output from the 7.8-MHz crystal filter will be a lower sideband signal.

This lower sideband signal is then amplified by TR6 and then fed to the transmit mixer along with the 35.2075-MHz VCO signal. The difference of these frequencies appears at the output of the transmit mixer. This difference frequency ranges from 27.4053 MHz to 27.408 MHz. This is the proper USB signal for audio modulating frequency range of 300 to 3000 Hz. The 7-MHz lower sideband signal is converted to a 7-MHz upper sideband signal in the transmit mixer. Table 7-7 traces the development of the USB transmit frequency for channel 40.

Lower Sideband Receive Mode

In the LSB receive mode, oscillator 3 is energized. Oscillator 3 operates at a frequency of 11.284166 MHz. This frequency is then tripled by the tripler (TR40) to become 3×11.284166 MHz, or 33.8525 MHz. The divisor N is 135. The VCO frequency can be found from the formula: $fVCO = f \times 3 + N \times 10^{-2}$. In the formula, fVCO is the VCO frequency in MHz, $f \times 3$ is the frequency at the output of tripler TR40, and N is the divisor of the programmable divider. Substituting, $fVCO = 33.8525 + 135 \times 10^{-2} = 33.8525 + 1.35$, or 35.2025 MHz.

This 35.2025-MHz VCO signal is fed to the receive mixer. There, it is heterodyned with the incoming signal. Suppose that there is a channel 40 LSB signal on the air being modulated by a 300 to 3000 Hz audio signal. This signal will be amplified and fed to the receive mixer. Here, it is mixed with the 35.2025-MHz VCO signal to produce a difference frequency which ranges from 7.7978 MHz to 7.8005 MHz. Thus, the 27-MHz *lower* sideband signal is converted by the mixer to a 7-MHz *upper* sideband signal. This signal is passed by the 7.8-MHz crystal filters and then amplified

Table 7-7. USB Transmit Mode Channel 40 Frequency Development.

Conditions: Channel 40 USB mode modulated by 300 Hz to 3000 Hz audio signal.

BALANCED MODULATOR

	Carrier Oscillator----------	7.802500 MHz	7.802500 MHz
	Audio Modulating Freq.-	+ 300 Hz to	+ 3000 Hz
(Sum)	Output of Balanced Mod-	7.802800 MHz to	7.805500 MHz
			(This is eliminated by the crystal filter)
	Carrier Oscillator----------	7.802500 MHz	7.802500 MHz
	Audio modulating Freq.--	- 300 Hz to	- 3000 Hz
(Difference)	Output of Balanced Mod-	7.802200 MHz to	7.799500 MHz
			(This is passed by the crystal filter)

TRANSMIT MIXER

VCO Signal-----------------	35.207500 MHz	35.207500 MHz
Xtal Filter out--------------	7.802200 MHz to	7.799500 MHz
Tx Mixer out---------------	27.405300 MHz to	27.408000 MHz

and fed to the SSB product detector. The difference of these two frequencies appears atr the output of the product detector. This frequency ranges from 300 to 3000 Hz, which is the same as the original modulating frequencies. Thus, the audio is recovered. Table 7-8 shows how the channel 40 LSB signal is processed in the receive mode.

Lower Sideband Transmit Mode

In the LSB transmit mode, oscillator 3 is again energized just as in the LSB receive mode. The VCO frequency will therefore be the same as for the LSB receive mode, 35.2025 MHz. This 35.2025-MHz VCO signal is fed to the transmit mixer stage, FET 7.

In the LSB transmit mode, the carrier oscillator will operate at 7.7975 MHz. This signal is buffered by TR28 and then fed to balanced modulator IC2. Also, the audio modulating signal is fed to the balanced modulator. With no audio input to the balanced modulator, there will be no output. If an audio signal is fed to the balanced modulator, the output from it will be a double sideband,

Table 7-8. LSB Receive Mode Frequency Development.

LSB Rx MODE FREQUENCY DEVELOPMENT

Conditions: Receiving channel 40 LSB signal modulated by 300 Hz to 3000 Hz audio signal.

RX MIXER

VCO Freq.----------------	35.202500 MHz	35.202500 MHz
Rx Signal------------------	27.404700 MHz to	27.402000 MHz
IF Signal------------------	7.797800 MHz to	7.800500 MHz

SSB PRODUCT DETECTOR

IF Signal------------------	7.797800 MHz to	7.800500 MHz
Carrier osc.----------------	7.797500 MHz	7.797500 MHz
Detected audio	300 Hz to	3000 Hz

suppressed-carrier signal. The upper sideband signal will be 7.7975 MHz plus the modulating frequency. The lower sideband signal will be 7.7975 MHz minus the modulating frequency.

If the audio modulating frequency ranges from 300 to 3000 Hz, the upper sideband signal will range from 7.7978 MHz to 7.8005 MHz. The lower sideband signal will range from 7.7972 MHz to 7.7945 MHz. These two sidebands are fed to the 7.8 MHz crystal filter. Only the upper sideband passes through the filter. The lower sideband is sharply attenuated because it does not fall within the passband of the filter.

The upper sideband signal from the output of the 7.8-MHz filter is amplified by TR6 and then fed to the transmit mixer along with the 35.2025-MHz VCO signal. The difference of these frequencies appears at the output of the transmit mixer. This difference frequency ranges from 27.4047 MHz to 27.402 MHz. This is the proper LSB, channel 40 signal for an audio modulating frequency range of 300 to 3000 Hz.

The 7-MHz upper sideband signal is converted to a 27-MHz lower sideband signal in the transmit mixer. Table 7-9 traces the development of LSB transmit frequency for channel 40. Table 7-10 shows the various oscillator frequencies used in the various modes of operation.

Table 7-9. LSB Transmit Mode Channel 40 Frequency Development.

LSB Tx MODE FREQUENCY DEVELOPMENT

Conditions: Channel 40 LSB mode modulated by 300 Hz to 3000 Hz audio signal.

BALANCED MODULATOR

	Carrier Oscillator----------	7.797500 MHz	7.797500 MHz
	Audio Modulating Freq.--	+ 300 Hz to	+ 3000 Hz
(Sum)	Output of Balanced Mod.-	7.797800 MHz to	7.800500 MHz
			(This is passed by the crystal filter.)
	Carrier Oscillator----------	7.797500 MHz	7.797500 MHz
	Audio modulating Freq.--	- 300 Hz to	- 3000 Hz
(Difference)	Output of Balanced Mod.-	7.797200 MHz to	7.794500 MHz
			(This is eliminated by the crystal filter.)

TRANSMIT MIXER

VCO Signal------------------	35.202500 MHz	35.202500 MHz
Xtal Filter Out	7.797800 MHz to	7.800500 MHz
Tx Mixer Out---------------	27.404700 MHz to	27.402000 MHz

Table 7-10. Oscillator Use in Various Modes of Operation.

MODE	Carrier Oscillator 7.7975	7.8025	Offset Oscillator #1-11.285	#2-11.2858	#3-11.2842
AM Rx			✓		
AM Tx		✓		✓	
USB Tx		✓		✓	
USB Tx		✓		✓	
LSB Rx	✓				✓
LSB Tx	✓				✓

Alignment

The alignment procedure for the President Model Madison is shown in Table 7-11. Use it in troubleshooting too.

Table 7-11. President Model Madison Alignment Procedure.

Step	Preset Condition	Connections	Adjustment	Remarks
1	Ch. 1. AM-Rx Mode Claritier in mid-position	R.F. Volt meter to secondary of L 24 (TP6)	L24	Adjust L24 for maximum indication on meter.
2	Same as step 1	DC voltmeter to pin 5 of IC 5 (TP7)	L17	Adjust L17 to obtain approx. 2.0 volts
3	Ch. 19. AM-RX Mode	R.F. voltmeter to secondary of L16 (TP8)	L16	Adjust L16 for maximum indication on meter.
4	Same as step 3	Frequency counter to TP8	CT6	Adjust CT6 for 34.9850 MHz +or -20 Hz.
5	Ch. 19. USB-Rx mode	Same as step 4	CT4	Adjust CT4 for 34.9875 MHz +or -20 Hz.
6	Ch. 19. LSB-Rx mode	Same as step 4	CT5	Adjust CT5 for 38.9825 MHz +or -20 Hz.
7	Ch. 19. USB-Tx mode	Same as step 4	VR9	Adjust VR9 for 34.9825 MHz +or -20 Hz.
8	Ch. 19. USB-Rx mode	Frequency counter to TP9 (emitter of TR28)	CT2	Adjust CT2 for 7.8025 MHz +5 Hz. -0 Hz.
9	Ch. 19. LSB-Rx mode	Same as step 8	CT3	Adjust CT3 for 7.7975 MHz +0 Hz. -5 Hz.

Troubleshooting

The following is a list of the more commonly encountered PLL trouble symptoms. Most of the symptoms will be caused by a missing VCO signal at TP8 (the secondary of L16) or by an incorrect VCO frequency at TP8. Anytime trouble is encountered with a PLL circuit, it is important to first check the lock detector output and the control voltage to the VCO control line. The lock detector output of the muPD858C is pin 1. An appropriate place to check the VCO control voltage is at TP7, pin 5 of IC5. The voltage level at these two points can tell you a lot about the condition of the loop.

No Receive or Transmit on All Channels and All Modes. If the PLL synthesizer circuit is the cause of this symptom, the VCO signal at TP8 will be missing or will be far off frequency. Check the lock detector output at pin 1 of IC7. If the lock detector output is high, several possibilities are indicated. Check for a missing 10.240-MHz reference frequency at pin 10 of IC7. Also, use a frequency counter to check the accuracy of the 10.240-MHz signal at pin 10. If the reference signal is missing, the 10.240-MHz crystal might be defective or the on-chip oscillator might be defective. If the reference signal at pin 10 is OK, check for the offset frequency signal at pin 11 of IC7. If this offset frequency is missing, use an rf voltmeter to trace the signal. Start at pin 11 of IC7 and trace through the PLL amps (TR30 and TR31), the PLL mixer (FET4), the tripler (TR40), and the oscillator circuitry. It is unlikely that this particular symptom would be caused by a defective 11-MHz oscillator because all three oscillators would have to be inoperative to affect all modes. If the offset frequency is present at pin 11, check the VCO control voltage at TP7 (pin 5 of IC5). If this voltage is abnormally high or low, the VCO frequency-determining circuit might be at fault. This could mean a defective IC5. Possibly, L17 has been badly mistuned or is defective. If L17 has been mistuned, you can correct this by the following method. Use a DC voltmeter to monitor the control voltage at TP7. While carefully tuning L17, watch for any change on the voltmeter. It is important to note the original setting of the tuning slug in L17 so that it can be returned there if tuning has no effect. If the voltage at TP7 begins to change while tuning L17, adjust for 2 volts on channel 1 (AM mode). If tuning L17 has no effect on the voltage at TP7, return the core of L17 to its original setting. Check for trouble on the VCO control line. Check bandpass filter (TR35) and the active filter which is part of IC7.

If the preceding tests didn't lead you to the cause of the trouble, maybe IC7 is defective. Before replacing it, though, make thorough tests on it. The μ PD858C has some very convenient test points for checking the operation of the IC. For example, pin 24 is the output of the programmable divider. The frequency at this terminal should be equal to the frequency at its input (pin 11) the N divisor programmed into the divider. If there is an input signal at pin 11 but not output signal at pin 24, the divider is probably defective. Or if the output signal from pin 24 is not equal to the input ÷ the N divisor, the divider is defective. In this case, replacement of the muPD858C would be necessary. You can also check the output of the reference divider at pin 6 of IC7. The frequency should be 10 kHz. If the frequency is wrong or missing, the IC is defective.

No Receive, AM Mode Only. If you look at Table 7-10, you will see that the No. 1 offset oscillator (11.285 MHz) is active only in the AM receive mode. This oscillator is the most likely suspect. Check oscillator crystal, supply voltage, transistor, etc. for any possible defect.

AM Receive Only. First check the lock detector output at pin 1 of IC7. If the voltage here is high, check for defective offset oscillator 2 (11.2858 MHz). If the voltage here is low, check for a missing 7.8025-MHz signal at TP9.

No LSB Receive or Transmit. First check the lock detector output at pin 1 of IC7. If the voltage here is high, check for defective offset oscillator 3 (11.2842 MHz). If the voltage here is low, check for a missing 7.7975-MHz signal at TP9.

Incorrect Synthesizer Output Frequency on All Modes and all Channels. If the synthesizer output frequency at TP8 is incorrect for all modes and all channels, either the reference oscillator (10.240 MHz) or all three 11-MHz offset oscillators must be operating off frequency. And if all three of the 11-MHz offset oscillators are operating off frequency, transmit the frequency control circuit might be at fault. It could also be a misadjusted VR9 or a bad zener diode, D31. It is possible—but less likely—that the reference oscillator is operating off frequency.

It takes a very large change in reference oscillator frequency to cause a small change in VCO frequency at TP8. For instance, suppose the reference oscillator is operating 5000 Hz off frequency at 10.245 MHz. Using channel 1 as an example, how much affect does this have on the VCO frequency at TP8. First, the 10.245-MHz signal is divided by 1024 to produce a reference frequency of 10,004.9 Hz. The input to the programmable divider (channel 1) will be 91 × 10,004.9, or 0.910446 MHz. Normally the input to the programmable divider would be 0.910000 MHz, so the offset frequency has changed by 446 Hz. The VCO frequency will change the same amount as the offset frequency. Actually, the change in offset frequency is a result of a change in VCO frequency. The VCO frequency, therefore, only changes 446 Hz, while the reference oscillator changed by 5000 Hz. The exact amount of VCO frequency change will vary, depending upon the channel because of the changing N divisors. For example, on channel 40 the VCO will change about 662 Hz as compared to about 446 on channel 1.

Incorrect Synthesizer Output Frequency on AM Receive Mode Only. If the VCO frequency at TP8 is wrong in the AM receive mode, but correct in all other modes, look for the 11.285-MHz oscillator (offset 1) to be operating off frequency. Adjust CT6 for proper frequency at TP8. If this can't be adjusted correct frequency, suspect a defective crystal (X6) or other components in the resonating circuit.

Incorrect Synthesizer Output Frequency on AM Transmit USB Receive and USB Transmit Modes. If the VCO frequency at TP8 is wrong in these modes, but OK in the other modes, look for the 11.2858-MHz oscillator (offset 2) to be operating off frequency. Adjust CT4 for the proper frequency at TP8. If this can't be adjusted to the correct frequency, suspect a defective crystal (X3) or other defective component in the resonating circuit.

Incorrect Synthesizer Output Frequency on LSB Receive and Transmit Modes. If the VCO frequency at TP8 is wrong in these modes but OK in all other modes, look for the 11.2842-MHz oscillator (offset 3) to be operating off frequency. Adjust CT5 for the proper frequency at TP8. If this can't be adjusted to correct frequency, suspect a defective crystal (X4) or other defective component in the resonating circuit.

Incorrect Transmit Frequency on AM and USB Modes. Notice that this is incorrect *transit frequency*, not *incorrect VCO* frequency. If the VCO frequency is OK, but the transmit frequency is incorrect in the AM and USB modes, check the frequency of the 7.8025-MHz AM-USB carrier signal at TP9. Adjust CT2 for exactly 7.8025 MHz at TP9. This will affect the demodulation of the USB receive signal as well.

Incorrect Transmit Frequency on LSB Mode. Again, notice that this is *incorrect transmit frequency,* not *incorrect VCO* frequency. If the VCO frequency at TP8 is correct but the transmit frequency is incorrect in the LSB mode, check the frequency of the 7.7975-MHz carrier oscillator signal at TP9. Adjust CT3 for exactly 7.7975 MHz at TP9. This will affect the demodulation of the LSB receive signal as well.

Incorrect Synthesizer Output Frequency on High Channels. If the VCO frequency at TP8 is correct for the low channels but incorrect on the higher channels, suspect a mistuned VCO. If the control voltage at TP7 is abnormally high, carefully tune L17 for normal voltage on one of the working channels.

Incorrect Synthesizer Output Frequency on Low Channels. If the VCO frequency at TP8 is correct for the high channels but incorrect for the lower channels, suspect a mistuned VCO. If the control voltage at TP7 is abnormally low, carefully tune L17 for normal voltage on one of the working channels.

Some Channels Inoperative. Check for proper binary levels on the program pins. Possibly, there is a defective channel switch or bad IC7. Also, check for possible open program lines due to printed circuit board cracks or those shorted across one or more lines because of component failure or solder bridge, etc.

Some Channels Intermittent. This is usually caused by a defective or dirty channel selector switch. Clean or replace as necessary.

Chapter 8 The µPD861C PLL IC

The µPD861C is a CMOS LSI device which is manufactured by NEC. The device is packaged in a 24-pin dual in-line package (DIP). As you will learn from the following discussion, this is a highly versatile device. The outline and pin identification of the µPD861C is shown in Fig. 8-1. Table 8-1 lists the absolute maximum ratings of the µPD861C. Table 8-2 lists the electrical characteristics of the µPD861C.

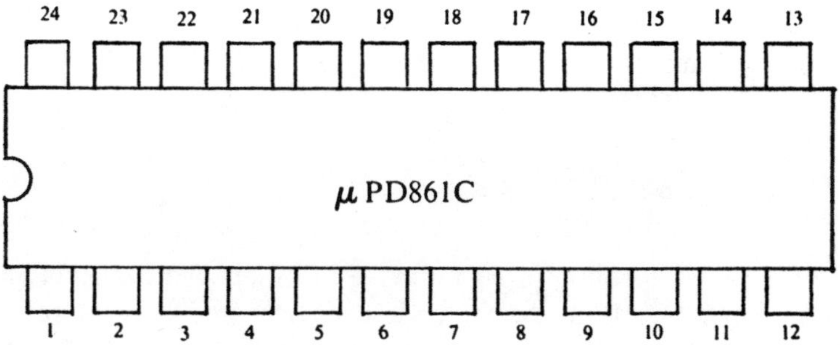

Fig. 8-1. An overview of the µPD861C IC, showing the pin numbering.

SPECIAL FEATURES

- ☐ On-chip reference oscillator (use external crystal)
- ☐ On-chip filter amplifier

☐ Code converter that simplifies programming

☐ Lock detector output for stopping transmitter if loop is unlocked

☐ BCD or pure binary programming capability

☐ Improper programming (because of defective or mispositioned channel selector) produces high level at pin 24

☐ High-speed and low-power consumption due to CMOS

☐ Single power supply and TTL compatible

☐ Pulldown resistors on program and mode switch inputs

☐ Wide operating temperature range (-35° C to +75° C)

☐ Output of reference divider and programmable divider brought out to external pins (this aids in troubleshooting)

Table 8-1. Absolute Maximum Ratings of the μPD861C (courtesy of NEC).

Ta (Ambient Temperature) = 25° C

Supply Voltage	VDD	-0.3 to +6.0	V.
Input voltage	Vi	-0.3 to +6.0	V.
Operating Temperature	Topt	-35 to +75	°C
Storage Temperature	Tstg	-55 to +125	°C

INTERNAL STRUCTURE

The basic structure of the μPD861C is shown in Fig. 8-2. Notice that all programming inputs (P1 through P8) connect to a *code converter*. Between the code converter and the programmable divider inputs is an *input mode switch*. This mode switch determines whether the programming is done in the BCD (offset by 90) or the pure binary form. If the mode switch is set for BCD (offset by 90), the channel number itself is programmed into the program lines, P1 through P8. The channel number is programmed in BCD form. The code converter accepts this BCD-coded channel number and then programs the programmable divider for the appropriate N number. Table 8-3 should help clari-

Table 8-2. Electrical Characteristics of the μPD861C (courtesy of NEC).

Characteristic	Symbol	Min.	Typ.	Max.	Unit	Test Conditions
Power supply voltage	VDD	4.5	5.0	5.5	V	
Total current	IDD			10	MA	f = 0
High level input voltage	VIH	0.8VDD		VDD	V	all inputs
Low level input voltage	VIL	-0.3		0.2VDD	V	all inputs
High level output voltage	VOH	0.85VDD		VDD	V	all outputs
Low level output voltage	VOL	0.0		0.15VDD		all outputs
Leak current	IL		1.0		nA	EO floating, AI, Ta = 25° C
Input Capacitance	Ci			10	pF	PI, FD, EP, XI, Vi = 0
Maximum frequency response	fd max	11.0			MHz	XI-X2, Divider
	fp max	3.0			MHz	Programmable • Divider

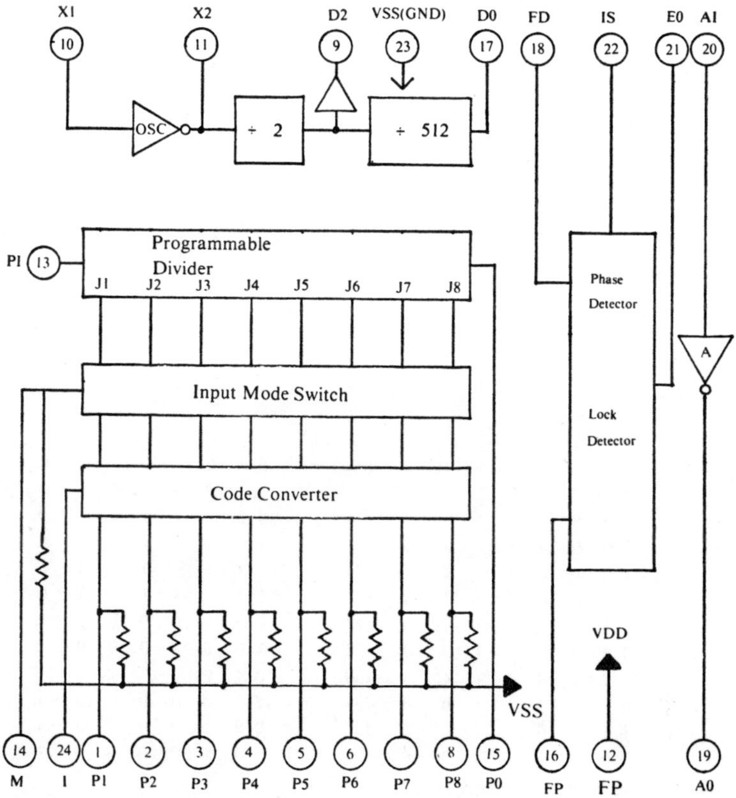

Fig. 8-2. Block diagram of the internal structure of the μPD861C IC (courtesy of NEC).

Table 8-3. Code Conversion for the μPD861C (courtesy of NEC).

CH.	PRG. Inputs								N	PRG Div. Input							
	BIN wt.	1	2	4	8	16	32	64	128	Weight 1	2	4	8	16	32	64	128
	BCD wt	1	2	4	8	10	20	40	80								
		P1	P2	P3	P4	P5	P6	P7	P8	J1	J2	J3	J4	J5	J6	J7	J8
1		1	0	0	0	0	0	0	0	91	1	1	0	1	1	0	1	0
2		0	1	0	0	0	0	0	0	92	0	0	1	1	1	0	1	0
3		1	1	0	0	0	0	0	0	93	1	0	1	1	1	0	1	0
4		0	0	1	0	0	0	0	0	95	1	1	1	1	1	0	1	0
5		1	0	1	0	0	0	0	0	96	0	0	0	0	0	1	1	0
6		0	1	1	0	0	0	0	0	97	1	0	0	0	0	1	1	0
7		1	1	1	0	0	0	0	0	98	0	1	0	0	0	1	1	0
8		0	0	0	1	0	0	0	0	100	0	0	1	0	0	1	1	0
9		1	0	0	1	0	0	0	0	101	1	0	1	0	0	1	1	0
10		0	0	0	0	1	0	0	0	102	0	1	1	0	0	1	1	0
11		1	0	0	0	1	0	0	0	103	1	1	1	0	0	1	1	0
12		0	1	0	0	1	0	0	0	105	1	0	0	1	0	1	1	0
13		1	1	0	0	1	0	0	0	106	0	1	0	1	0	1	1	0
14		0	0	1	0	1	0	0	0	107	1	1	0	1	0	1	1	0
15		1	0	1	0	1	0	0	0	108	0	0	1	1	0	1	1	0
16		0	1	1	0	1	0	0	0	110	0	1	1	1	0	1	1	0
17		1	1	1	0	1	0	0	0	111	1	1	1	1	0	1	1	0
18		0	0	0	1	1	0	0	0	112	0	0	0	0	1	1	1	0
19		1	0	0	1	1	0	0	0	113	1	0	0	0	1	1	1	0
20		0	0	0	0	0	1	0	0	115	1	1	0	0	1	1	1	0
21		1	0	0	0	0	1	0	0	116	0	0	1	0	1	1	1	0
22		0	1	0	0	0	1	0	0	117	1	0	1	0	1	1	1	0
23		1	1	0	0	0	1	0	0	120	0	0	0	1	1	1	1	0
24		0	0	1	0	0	1	0	0	118	0	1	1	0	1	1	1	0
25		1	0	1	0	0	1	0	0	119	1	1	1	0	1	1	1	0
26		0	1	1	0	0	1	0	0	121	1	0	0	1	1	1	1	0
27		1	1	1	0	0	1	0	0	122	0	1	0	1	1	1	1	0
28		0	0	0	1	0	1	0	0	123	1	1	0	1	1	1	1	0
29		1	0	0	1	0	1	0	0	124	0	0	1	1	1	1	1	0
30		0	0	0	0	1	1	0	0	125	1	0	1	1	1	1	1	0
31		1	0	0	0	1	1	0	0	126	0	1	1	1	1	1	1	0
32		0	1	0	0	1	1	0	0	127	1	1	1	1	1	1	1	0
33		1	1	0	0	1	1	0	0	128	0	0	0	0	0	0	0	1
34		0	0	1	0	1	1	0	0	129	1	0	0	0	0	0	0	1
35		1	0	1	0	1	1	0	0	130	0	1	0	0	0	0	0	1
36		0	1	1	0	1	1	0	0	131	1	1	0	0	0	0	0	1
37		1	1	1	0	1	1	0	0	132	0	0	1	0	0	0	0	1
38		0	0	0	1	1	1	0	0	133	1	0	1	0	0	0	0	1
39		1	0	0	1	1	1	0	0	134	0	1	1	0	0	0	0	1
40		0	0	0	0	0	0	1	0	135	1	1	1	0	0	0	0	1
40		0	0	0	0	0	0	0	0	135	1	1	1	0	0	0	0	1

fy this for you. Take channel 15, for example. The BCD code for 15 is shown. Notice that the divisor N is 108, but it was stated that the BCD was offset by 90. And 90 + 15 = 105, not 108. Where did the N value of 108 come from? If you will look at Table 8-3, you will notice that on channels 1 through 3 the N divisor is equal to the channel number + 90. But after channel 3, this doesn't hold true. This is because there is a 20-kHz gap between channels 3 and 4, whereas there was only a 10-kHz gap between channels 1 and 2 and 2 and 3. There is also a 20-kHz gap between channels 7 and 8 and 11 and 12. These three gaps are the reason that the N number for channel 15 is 108 instead of 105. The divisor N has to change according to channel spacing so the code converter is designed to produce the appropriate N divisor according to the channel spacing, not the channel number.

The input to the programmable divider itself will be the pure binary equivalent of the N number. For example, on channel 15, for an N divisor of 108 the appropriate binary levels are applied to J1 through J8 as shown. If you were to add the various weights where the binary 1 level appears, the sum would be equal to the N number. So the function of the code converter is simply to *translate* the BCD channel number into a pure binary code to be applied to the programming inputs (J1 through J8) of the programmable divider.

If the mode switch is set for pure binary coding the binary levels at P1 through P8 are passed on to programming inputs J1 through J8. Table 8-4 shows how various N numbers are programmed in either mode, BCD (offset by 90) or binary. The function of the various pins of the μPD861C is as follows:

- ☐ Pins 1 through 8. These are the program inputs to the programmable divider. As mentioned, the programming can be done either in BCD (offset by 90) or pure binary form. Pin 1 is the least significant bit, while pin 8 is the most significant bit. The weights of pins 1 through 4, are the same for either program mode, but the weights of pins 5 through 8 are different for the two modes. The weights of these pins are shown in Tables 8-3 and 8-4 for the two program modes. Each program line has an internal pulldown resistor on the IC chip.
- ☐ Pin 9. A signal is available at pin 9 which is equal to one-half of the reference oscillator frequency. With a 10.240-MHz reference oscillator, the frequency at pin 9 will be 5.12 MHz. This 5.12-MHz signal can be multiplied by three to yield 15.360 MHz. This 15.360-MHz signal can be mixed with a 16-MHz VCO signal to produce an offset frequency for the program-

Table 8-4. μPD861C Programming.

"M" (Pin 14)	"N"	P1 (1)	P2 (2)	P3 (4)	P4 (8)	P5 (16)	P6 (32)	P7 (64)	P8 (128)	
	3	1	1	0	0	0	0	0	0	
	4	0	0	1	0	0	0	0	0	
Lo	5	1	0	1	0	0	0	0	0	
	.									
	.									
	.									
	91	1	1	0	1	1	0	1	0	
	92	0	0	1	1	1	0	1	0	
	93	1	0	1	1	1	0	1	0	
	94	0	1	1	1	1	0	1	0	
	95	1	1	1	1	1	0	1	0	
	96	0	0	0	0	0	1	1	0	
	.									
	.									
	.									Binary coding
	250	0	1	0	1	1	1	1	1	
	251	1	1	0	1	1	1	1	1	
	252	0	0	1	1	1	1	1	1	
	253	1	0	1	1	1	1	1	1	
	254	0	1	1	1	1	1	1	1	
	255	1	1	1	1	1	1	1	1	
	0	0	0	0	0	0	0	0	0	
	1	1	0	0	0	0	0	0	0	Program inhibit
	2	0	1	0	0	0	0	0	0	
		P1	P2	P3	P4	P5	P6	P7	P8	
	CH. #	(1)	(2)	(4)	(8)	(10)	(20)	(40)	(80)	
	1	1	0	0	0	0	0	0	0	
Hi	2	0	1	0	0	0	0	0	0	
	3	1	1	0	0	0	0	0	0	
	4	0	0	1	0	0	0	0	0	
	.									
	.									
	.									
	10	0	0	0	0	1	0	0	0	
	11	1	0	0	0	1	0	0	0	
	12	0	1	0	0	1	0	0	0	
	13	1	1	0	0	1	0	0	0	
	.									
	.									
	.									BCD coding
	38	0	0	0	1	1	1	0	0	
	39	1	0	0	1	1	1	0	0	
	40	0	0	0	0	0	0	1	0	
	40	0	0	0	0	0	0	0	0	
N 40	0	1	0	1						
		1	1	0	1					
		0	0	1	1					
		1	0	1	1					
		0	1	1	1					
		1	1	1	1					Program inhibit

mable divider. With the BCD (offset by 90) program mode, this works out perfectly for the available N values (not by accident). For example, with a 16.270-MHz VCO frequency, the offset frequency would be 16.270 MHz - 15.360 MHz, or 0.91 MHz (91 being the exact N value for channel 1).

□ Pins 10 and 11. The reference oscillator crystal (usually 10.240 MHz) is connected between these two pins. Pin 10 is the input to an on-chip amplifier, while pin 11 is the output. The crystal between the output and the input of the amplifier determines the resonant frequency. Bias is supplied by a resistor connected between pins 10 & 11.

□ Pin 12. This is the voltage supply point for the IC. The normal supply voltage is + 5 volts.

□ Pin 13. This is the input terminal of the programmable divider. The input frequency here should be held to 3 MHz or less.

□ Pin 14. This is the program input mode switch. If pin 14 is grounded or just left open, the N divisor will be determined by applying pure binary programming to program lines P1 through P8. If pin 14 is connected to the supply voltage (VDD), the programming will be done in the BCD form. The N value in this case will be BCD + 90 (plus any channel gaps, as explained). There is also a pulldown resistor on the IC chip for this terminal.

□ Pin 15. This is the output of the programmable divider. This is a very convenient test point. It allows you to determine if the divider is working properly. This terminal pin is connected to one of the phase detector inputs through an external connection.

□ Pin 16. This is a phase detector input.

□ Pin 17. This is the output of the reference divider, another convenient test point. This serves as the reference frequency, 10 kHz.

□ Pin 18. This is the other phase detector input.

□ Pin 19. This is the output terminal of the active filter amplifier.

□ Pin 20. This is the input of the active filter amplifier.

□ Pin 21. This is an error signal output of the phase detector.

□ Pin 22. When the loop is locked, a HIGH level signal exists at pin 22. When the loop falls out of lock, a pulsed wave appears at pin 22 which is filtered and used to inhibit transmitter output.

□ Pin 23. This is the ground or VSS point for the IC.

□ Pin 24. This is the inhibit output. If improper programming is applied to P1 through P8, the voltage at pin 24 goes to a HIGH level. Under normal programming, the voltage level here is LOW. This can be used to stop the transmitter in case of *false* programming.

PRACTICAL APPLICATION

This particular application of the μPD861C is used in the Sears model 934.38081700 CB mobile (Road Talker 40) transceiver. The block diagram of the PLL portion of the set is shown in Fig. 8-3. Pin 14 is connected to + 5 volts (VDD). This means that the programming is set for the BCD code. If you look at Table 8-3, you will find the various BCD codes listed for all 40 channels. Table 8-3 shows that channel 40 has *two* possible BCD codes that relate to the same N number, 135. In this set, for channel 40, all program lines (P1 through P8) are at binary 0.

The VCO operates in the 16 to 17 MHz range. Table 8-5 lists the VCO frequencies for the 40 channels. The 5.12-MHz signal from pin 9 of the μPD861C is fed to the primary of T303, which is tuned to the third harmonic of 5.12 MHz. This is 15.360 MHz. This 15.360-MHz signal is fed from the secondary of T303 to the down mixer along with the VCO signal. The output of the down mixer is equal to the VCO frequency minus 15.360 MHz. This offset frequency is then buffered and fed to the input of the programmable divider at pin 13.

Let's analyze the receive and transmit mode of operation for channel 1. When the channel selector is switched to channel 1, the programmable divider is set to divide by 91. If the output of the programmable divider is to be 10 kHz (which it must be in order for the loop to be locked), the input to the programmable divider must be 91 × 10 kHz, or 0.91 MHz. This then would mean that the VCO frequency would have to be 15.360 MHz + 10.91 MHz, or 16.270 MHz. In the receive mode, this 16.270-MHz signal is fed to the first mixer. Here, it is heterodyned with the incoming channel 1 frequency (26.965 MHz) to produce a difference frequency of 10.695 MHz. This 10.695-MHz signal is the first i-f. This 10.695-MHz first i-f is then fed to the second mixer. There, it is heterodyned with a signal from the 10.240-MHz reference oscillator to produce a difference frequency of 455 kHz. This 455-kHz signal is the second intermediate frequency.

In the transmit mode the VCO signal is fed to the transmit mixer along with a 10.695-MHz signal from the transmit oscillator. The output of the transmit mixer is the sum of the two frequencies: 16.270 MHz + 10.695 MHz = 26.965 MHz, the correct transmit frequency for channel 1. This signal is then fed to the

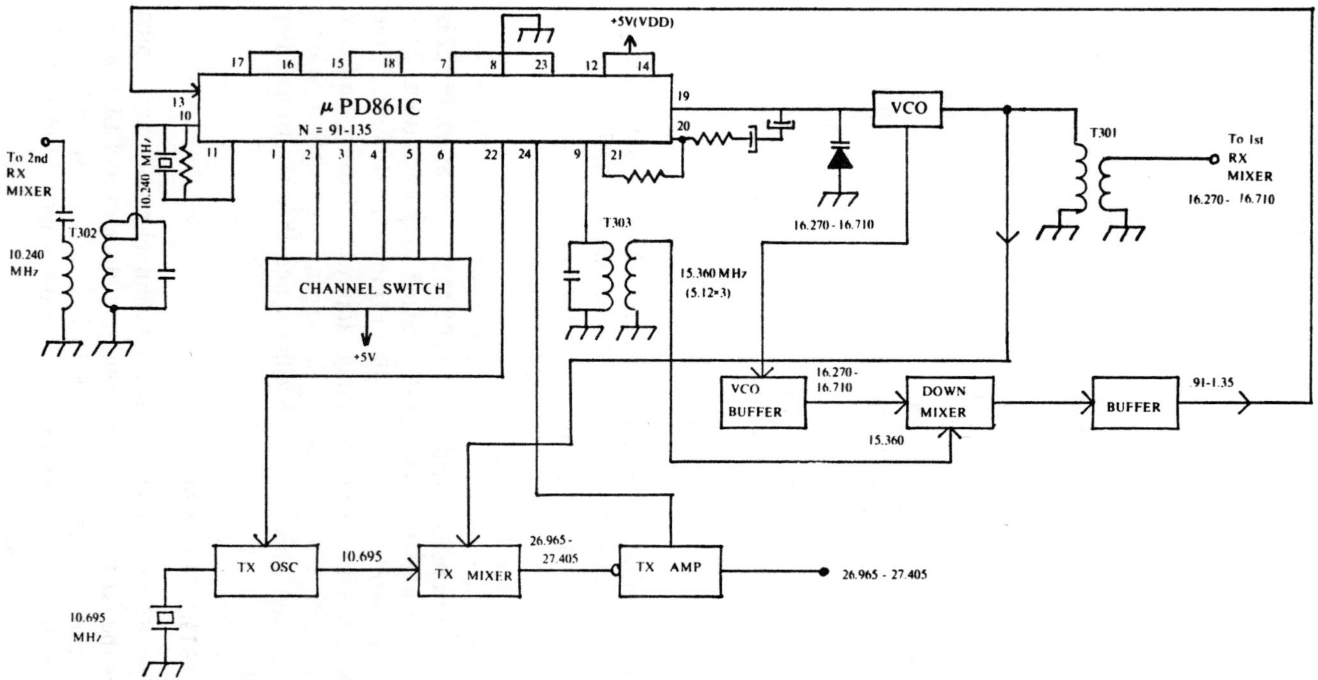

Fig. 8-3. Block diagram of the PLL synthesizer circuit used in the Sears 934.38081700 model (Road Talker 40) CB radio. The circuit uses the μPD861C in the BCD programming mode.

Table 8-5. Sears Road Talker 40 PLL VCO Frequencies.

Channel No.	VCO Frequency (MHz)	Channel No.	VCO Frequency (MHz)
1	16.270	21	16.520
2	16.280	22	16.530
3	16.290	23	16.560
4	16.310	24	16.540
5	16.320	25	16.550
6	16.330	26	16.570
7	16.340	27	16.580
8	16.360	28	16.590
9	16.370	29	16.600
10	16.380	30	16.610
11	16.390	31	16.620
12	16.410	32	16.630
13	16.420	33	16.640
14	16.430	34	16.650
15	16.440	35	16.660
16	16.460	36	16.670
17	16.470	37	16.680
18	16.480	38	16.690
19	16.490	39	16.700
20	16.510	40	16.710

transmitter amplifier and following stages. Notice that a line from pin 24 of μPD861C connects to the transmit amplifier stage. If improper programming is applied to the program lines, the voltage at pin 24 goes HIGH. This voltage is used to reverse bias the transmitter amplifier stage. This stops the transmitter output.

Also notice that a line from pin 22 of the μPD861C connects to the transmit oscillator (10.695 MHz oscillator). Pin 22 is normally at a HIGH level under locked loop conditions. This supplies forward bias to the oscillator transistor. However, if the loop becomes unlocked, the voltage at pin 22 will drop to a LOW level (pulsed). This then causes the transmit oscillator to lose its forward bias. This kills the oscillator and thus the transmitter output.

REALISTIC TRC-424

The Realistic TRC-424 is a 40-channel mobile unit which utilizes the μPD861C as the building block of its PLL synthesizer circuit. The schematic of the PLL circuit is shown in Fig. 8-4. The function and operation of the various stages of this PLL circuit are as follows:

PLL IC

From studying the IC801 pin connections, you can see that pin 14 is grounded. This means that the programming will be done in the pure binary form. Pins 7 and 8 connect directly to the +6-volt supply. The other program inputs (pins 1 through 6) connect to the channel selector switch. Because pins 7 and 8 (with respective weights of 64 and 128) are tied directly to the + 6-volt supply, the divisor N can never be less than 192 (64 + 128) even if all the other program inputs are at binary 0. Since the minimum number programmed by the channel selector is 4, this sets the minimum divisor used at 192 + 4, or 196. And since the maximum number programmed by the channel selector is 48, the maximum divisor used is 192 + 48, or 240. Notice on the program lines to the channel selector that external pulldown resistors are used. These are actually six resistors in a single package. The μPD861C also has on-chip pulldown resistors on the program inputs.

The programming code for each of the 40 channels is shown in Table 8-6. Also shown in this table are the VCO frequency and the N divisor of each channel.

Reference Oscillator

This PLL circuit utilizes the on-chip oscillator. The 10.240-MHz crystal is connected between pins 10 and 11. Transformer T801 is used to couple a portion of the 10.240-MHz signal to the receive second mixer. There is no adjustment provided to fine tune the frequency of the 10.240-MHz oscillator. This is really not needed anyway.

Voltage-Controlled Oscillator

The VCO oscillator circuit consists of transistor Q801, varactor diode D802, and associated components. This is basically a Hartley oscillator utilizing an FET transistor. The primary of transformer T802, along with capacitors C815, C817, C814, and varactor D802, form the resonant circuit for the VCO. Course frequency adjustments are made by tuning T802. Once T802 is adjusted to place the oscillator frequency in the "ballpark," fine tuning of the VCO is accomplished by the bias on the VCO control line, which varies the varactor capacitance. The VCO frequency ranges from 36.750 MHz for channel 1 to 37.190 MHz for channel 40.

Associated with the VCO oscillator are two buffer stages.

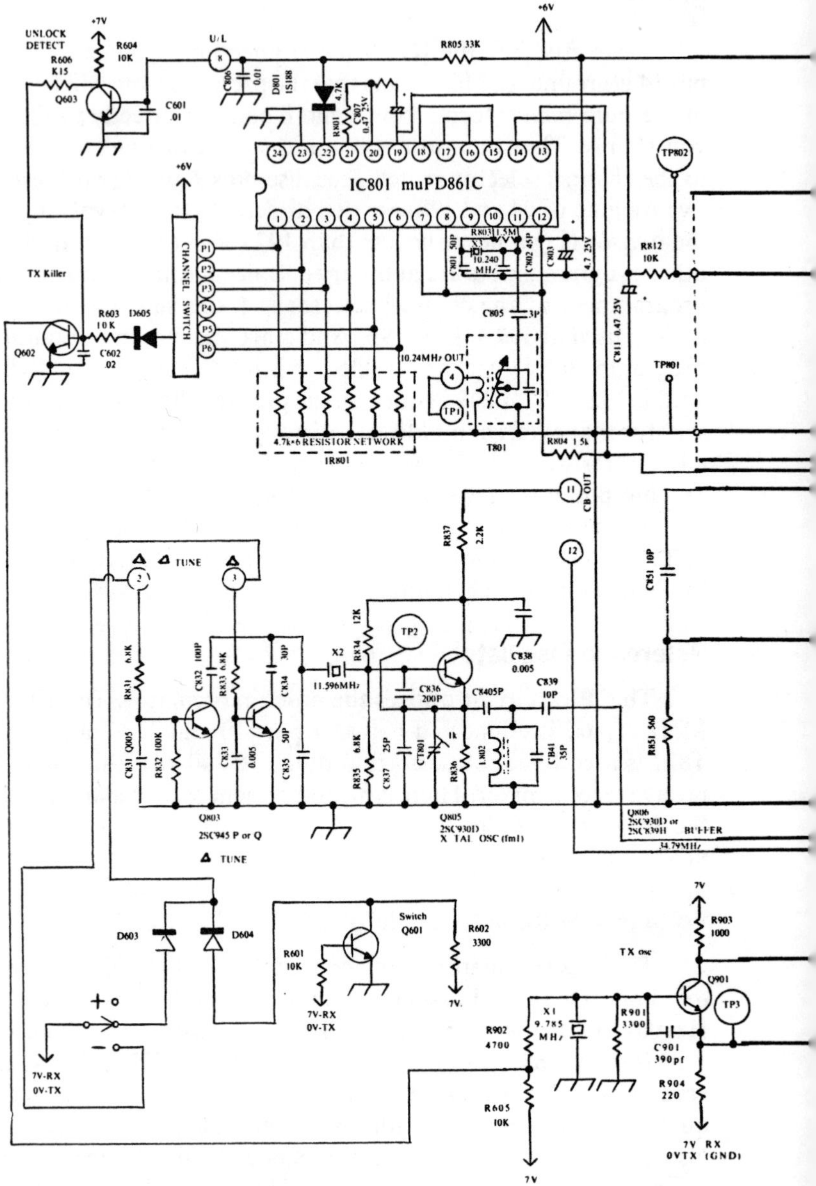

Fig. 8-4. Complete schematic of the PLL synthesizer used in the Realistic TRC-424 CB radio. This synthesizer uses the μPD861C in the pure binary mode (courtesy of Radio Shack).

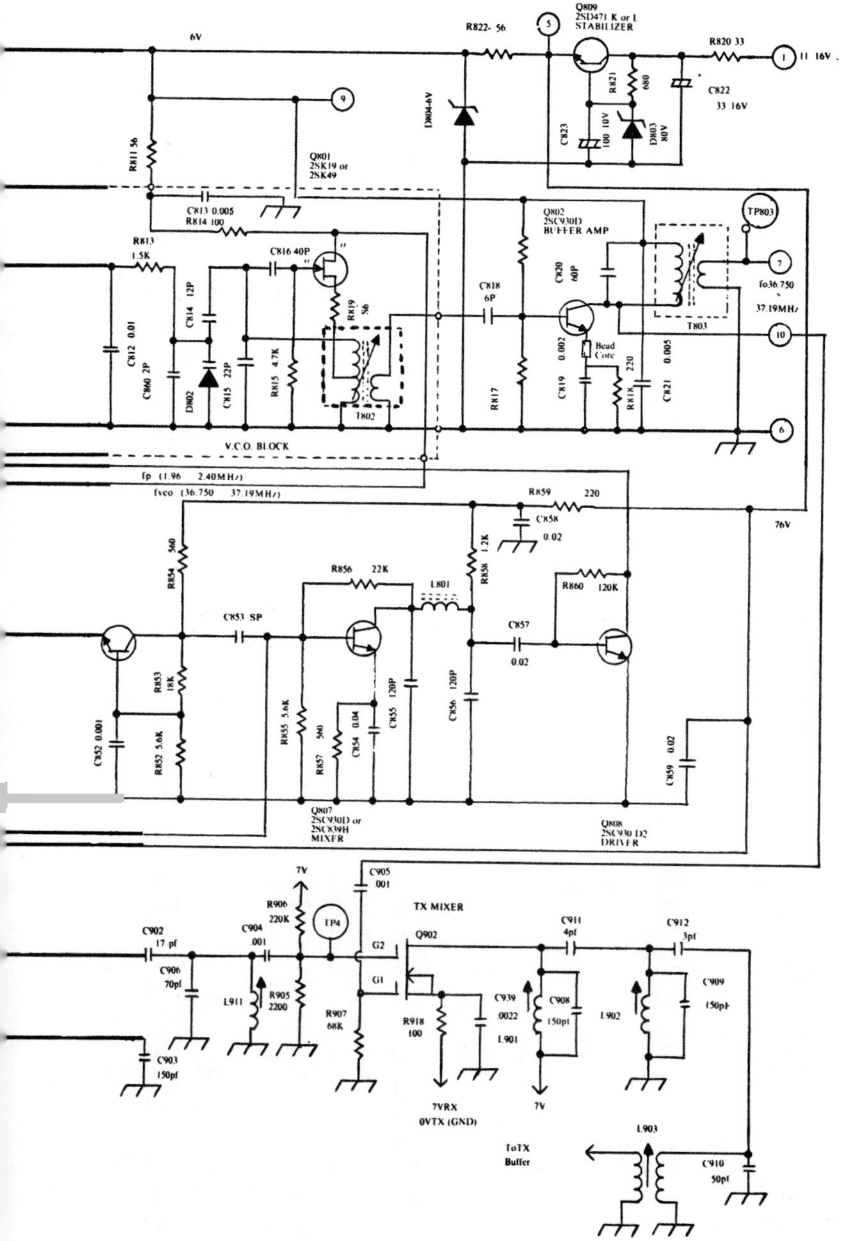

6V
R822- 56
Q809
2SD471 K or L
STABILIZER
R820 33
11 16V
R821 680
C822
33 16V
D804-6V
C823 100 10V
D803 80V
R811 56
Q801
2SK19 or
2SK49
C813 0.005
R814 100
Q802
2SC930D
BUFFER AMP
TP803
R813
1.5K
C816 40P
C814 12P
C812 0.01
C860 2P
D802
C815 22P
R815 4.7K
R819 56
T802
C818
6P
C820 60P
fo36.750
37.19MHz
T803
C819 0.002
Bead
Core
R818 220
C821 0.005
R817
V.C.O. BLOCK
fp (1.96 2.40MHz)
fvco (36.750 37.19MHz)
R859 220
C858
0.02
76V
R854 560
R858 1.2K
R856 22K
L801
R860 120K
C853 5P
C857
0.02
R853 18K
C852 0.001
R852 5.6K
R855 5.6K
R857 560
C854 0.04
C855 120P
C856 120P
C859 0.02
Q807
2SC930D or
2SC839H
MIXER
Q808
2SC930 D2
DRIVER
7V
C905
.001
TX MIXER
R906
220K
TP4
C902
17 pf
C904
.001
Q902
G2
G1
C911
4pf
C912
3pf
C906
70pf
L911
R905
2200
R907
68K
R918
100
C939
.0022
C908
150pf
L901
L902
C909
150pF
C903
150pf
7VRX
0VTX (GND)
7V
L903
To TX
Buffer
C910
50pf

Table 8-6. N Numbers and VCO Frequencies.

Channel No.	"N" Divisor	Prog. code (pin weight)-1	2	4	8	16	32	VCO Frequency (MHz)
		(pin)----------1	2	3	4	5	6	
1	196	0	0	1	0	0	0	36.750
2	197	1	0	1	0	0	0	36.760
3	198	0	1	1	0	0	0	36.770
4	200	0	0	0	1	0	0	36.790
5	201	1	0	0	1	0	0	36.800
6	202	0	1	0	1	0	0	36.810
7	203	1	1	0	1	0	0	36.820
8	205	1	0	1	1	0	0	36.840
9	206	0	1	1	1	0	0	36.850
10	207	1	1	1	1	0	0	36.860
11	208	0	0	0	0	1	0	36.870
12	210	0	1	0	0	1	0	36.890
13	211	1	1	0	0	1	0	36.900
14	212	0	0	1	0	1	0	36.910
15	213	1	0	1	0	1	0	36.920
16	215	1	1	1	0	1	0	36.940
17	216	0	0	0	1	1	0	36.950
18	217	1	0	0	1	1	0	36.960
19	218	0	1	0	1	1	0	36.970
20	220	0	0	1	1	1	0	36.990
21	221	1	0	1	1	1	0	37.000
22	222	0	1	1	1	1	0	37.010
23	225	1	0	0	0	0	1	37.040
24	223	1	1	1	1	1	0	37.020
25	224	0	0	0	0	0	1	37.030
26	226	0	1	0	0	0	1	37.050
27	227	1	1	0	0	0	1	37.060
28	228	0	0	1	0	0	1	37.070
29	229	1	0	1	0	0	1	37.080
30	230	0	1	1	0	0	1	37.090
31	231	1	1	1	0	0	1	37.100
32	232	0	0	0	1	0	1	37.110
33	233	1	0	0	1	0	1	37.120
34	234	0	1	0	1	0	1	37.130
35	235	1	1	0	1	0	1	37.140
36	236	0	0	1	1	0	1	37.150
37	237	1	0	1	1	0	1	37.160
38	238	0	1	1	1	0	1	37.170
39	239	1	1	1	1	0	1	37.180
40	240	0	0	0	0	1	1	37.190

One of these buffers is Q802. This buffer stage amplifies the VCO signal and feeds it through T803 to the first receive mixer. Also from the collector of Q802, the VCO signal is fed to the transmit mixer.

The other buffer is Q806. This is connected in the common-base configuration. The VCO signal is fed from the drain of the VCO to the emitter of the buffer, where it is amplified and then fed to the loop mixer transistor.

Loop Oscillator/Delta-Tune

The loop oscillator (sometimes called down oscillator or offset oscillator) consists of Q805, crystal X2, and associated circuitry. The oscillator operates at a frequency of 11.5966 MHz, as governed by crystal X2. However, the tank circuit consisting of C841 and L802 is resonant at the third harmonic of 11.5966 MHz. This yields a frequency of 3 × 11.5966, or 34.79 MHz. This 34.79-MHz signal is fed to the loop mixer along with the VCO signal to produce an offset frequency for the programmable divider.

The delta-tune circuit consists of transistors Q804 and Q803, along with the associated capacitors C834 and C832. When the delta-tune switch (S2) is in the center position, transistor Q804 is forward biased through diode D603. This causes Q804 to saturate, thus grounding C834, the 30-pF capacitor that connects to one side of crystal X2. This sets the oscillator at its center frequency of 11.5966 MHz. If the delta-tune switch is switched to the "+" side, Q804 will lose its forward bias. This causes it to cut off. This action ungrounds capacitor C834, effectively removing it from the circuit. With less capacitance between the crystal and ground, the crystal operates at a higher frequency. If the delta-tune switch is placed in the "-" position, transistor Q803 is forward biased. This causes Q803 to saturate, thus grounding C832, the 100-pF capacitor that connects to one side of crystal X2. With this larger capacitance between X2 and ground, the crystal will operate at a lower frequency. The delta-tune control can increase or decrease the oscillator frequency by approximately 1.5 kHz. The VCO frequency will be forced to change in exact accordance with the oscillator frequency (multiplied by 3). Thus, off-frequency signals can be tuned in with better clarity. The delta-tune function is only operable in the receive mode. In the receive mode, transistor Q601 is biased to the point of saturation. This places its collector at ground potential (practically). Thus, D604 is reversed biased. In the transmit mode, Q601 loses its bias and the collector voltage rises to 5 volts or so. This forward biases diode D604 which in turn allows transistor Q804 to become forward biased. Also in the transmit mode, the supply voltage to the delta-tune switch (S2) is removed to prevent the switch from forward biasing Q803. Therefore, in the transmit mode, transistor Q804 is turned on and Q803 is turned off. This sets the oscillator to its center frequency. The center frequency of the oscillator can be precisely set by CT801.

Loop Mixer

The loop mixer (down mixer or offset mixer) is a simple common-emitter circuit. The loop mixer is fed two signals, one from the VCO buffer (Q806) and the other from the loop oscillator. The output of the mixer is the difference of these two frequencies. The loop oscillator frequency is a constant 34.79 MHz on all channels. The VCO frequency will range from 36.750 MHz for channel 1 to 37.190 MHz for channel 40. The output of the loop mixer will therefore range from 1.96 MHz (36.750 MHz minus 34.790 MHz) for channel 1 to 2.40 MHz (37.190 MHz minus 34.790 MHz) for channel 40. In the collector circuit of the mixer, a low-pass pi network filter exists. This passes only the difference frequency and sharply attenuates the other frequencies.

Loop Driver

This is a very simple common-emitter circuit used to increase the level of the offset frequency before it is applied to the input of the programmable divider.

Transmit Killer/Unlock Detector

The transmitter killer is a simple transistor switch. There are two inputs to the base of Q602, one from the channel selector switch through R603 and D605 and the other from the collector of Q603 (unlock detector) through R606. Let's first talk about the input from the channel selector. If a channel selector develops loose or dirty contacts, it is possible that the programming to the programmable divider could be such as to cause an improper or illegal frequency to be developed. Or if an operator placed the channel switch between channels, the same thing could occur. When this occurs, the voltage at the anode of D605 (at the channel selector switch) will go to a HIGH level.This HIGH level voltage forward biases Q602, causing it to switch on. When Q602 switches on, it kills the forward bias to the transmitter oscillator by grounding the junction of R902 and R605. Thus, the transmitter is disabled because the transmit oscillator can't operate without forward bias.

Now let's check the other input to the transmit killer. This comes from the collector of Q603, the lock detector. Normally, under locked loop conditions, the voltage level at pin 22 of IC801 is at a HIGH level (around + 6 volts). Under these conditions, the lock detector (Q603) will be forward biased through resistor R805 back to the + 6-volt supply. Thus, the collector of Q603 will be near ground potential. However, if the loop becomes unlocked, the voltage at pin 22 of IC801 falls to a LOW level pulsed. Diode D801 is then forward biased. This LOW level at pin 22 pulls down the voltage at the base of Q603, causing Q603 to turn off. When Q603 turns off, its collector rises to a HIGH level. This HIGH level voltage then forward biases the transmit killer, Q602 through R606. Thus the transmit killer turns on, killing the transmitter. The action of the transmit killer, Q602, can be compared with that of a two-input NOR gate. When either input is HIGH, the output is LOW. The output is HIGH only when both inputs are LOW.

Transmit Oscillator

The transmit oscillator operates at a frequency of 9.785 MHz. The oscillator is active only in the transmit mode. The frequency is controlled by crystal X1. The purpose of the transmit oscillator is to provide a signal to mix with the VCO frequency to produce the channel frequency.

Transmit Mixer

The transmit mixer is a dual-gate FET transistor. The VCO signal is fed to gate 1 of the FET, while the transmit oscillator signal is fed to gate 2 of the FET. The two signals are then heterodyned in the FET to produce a difference frequency at the output (drain) which is equal to the channel frequency. The tuned circuit in the drain of the FET is tuned to this 27-MHz difference frequency. The sum and original frequencies are attenuated.

Stabilizer Circuit

The stabilizer circuit is nothing but a voltage regulator. The 8-volt zener diode at the base of Q809 holds the base voltage constant. If the loading on the emitter increases, this increases the forward bias on the transistor causing it to conduct harder. This decreases the voltage drop across the emitter/collector, making up for the heavier loading on the emitter. If the loading on the emitter decreases, this decreases the forward bias on the transistor and causes the transistor conduction to decrease. This decreased conduction increases the voltage drop from the emitter/collector, making up for the decreased loading. The voltage at the emitter is kept at approximately 7 volts. Between the emitter and ground, there is a series circuit consisting of a 6-volt zener diode (D804) and a 56-ohm current-limiting resistor (R822). At the junction of the zener diode and the current limiting resistor, there is +6 volts. This is highly regulated due to both the 6-volt zener and the series pass regulator transistor. This highly regulated 6 volts then serves as VDD supply for the μPD861C IC.

Analysis of Operation

When the channel selector is set for channel 40, the programmable divider is programmed for an N divisor of 240. Table 8-6 shows the programming code on pins 1 through 6 for channel 40. Remember that programming pins 7 and 8 are connected directly to VDD so they are at binary 1, regardless of the channel selector position. The total N divisor is equal to 192 plus the N number programmed into pins 1-6. For channel 40, the N number programmed into pins 1 through 6 is 48 so the total N number is 192+ 48, or 240.

When the radio is first switched on, the VCO will start operating somewhere around 37 MHz. Say that the VCO first starts operating at exactly 37.000 MHz. This 37.000-MHz signal is sent to the loop mixer through VCO buffer Q806. Also, a 34.790-MHz signal from the loop oscillator is fed to the loop mixer. These two signals are then heterodyned in the mixer. The difference of the two frequencies appears in the output. This difference frequency is 37.000 MHz - 34.790 MHz, or 2.21 MHz. This frequency is passed by the low-pass filter to the driver stage, where the signal is amplified and fed to the input of the programmable divider at pin 13 of IC801. This 2.21-MHz signal is divided by 240 to yield a 9.208-kHz signal at the output of the programmable divider. This 9.208-kHz signal is then fed to the phase detector. Here, it is compared with the 10-kHz reference signal. The difference of the phase detector inputs causes an error signal to appear at the output of the phase detector. This error signal is then filtered by the filter amplifier and applied to the VCO control line at pin 19 of IC801. The VCO control line also has a filter network for further smoothing of the DC control voltage. The error signal causes the DC control voltage on the VCO control line to increase to a higher positive value. This increases the frequency of the VCO. As the frequency approaches 37.19 MHz, the input to the programmable divider approaches 2.40 MHz and the output of the programmable divider approaches the 10-kHz reference frequency. As this occurs, the error signal from the phase detector grows smaller, so that finally when the VCO frequency reaches 37.190 MHz, the error signal from the phase detector drops to a minimal level. The DC control voltage keeps the VCO on track.

In the receive mode, the 37.190-MHz VCO signal is sent to the first mixer. There, it is heterodyned with a 27.405-MHz, channel 40 signal to produce a difference frequency of 9.785 MHz. This is the first i-f. This 9.785-MHz first i-f is then fed to the second mixer, where it is heterodyned with a 10.240-MHz signal from the reference oscillator to produce a difference frequency of 455 kHz. This is the second intermediate frequency.

In the transmit mode, the 37.190-MHz VCO signal is fed to the transmit mixer, Q902, along with a 9.785-MHz signal from the transmit oscillator. These two frequencies are heterodyned to produce a difference frequency of 27.405 MHz. This is the correct channel 40 transmit frequency. It is then amplified by the following stages.

Alignment

☐ Use a frequency counter to verify that the 10.240-MHz reference oscillator frequency is correct. Check this at TP1, the secondary of T801. The frequency should be within + 500 Hz. There is no adjustment to correct this if it is wrong.

☐ Touch an rf voltmeter probe to TP1 and adjust T801 for maximum rf indication.

☐ Check the frequency of the 11.5966-MHz loop oscillator at TP2, the emitter of Q805. Use a 5-pF or 10-pF coupling capacitor with a counter probe to prevent capacitive loading of the oscillator; otherwise, the oscillator frequency might shift too much with the counter connected. Adjust CT801 for exactly 11.5966 MHz. Make certain that the delta-tune control is in the "C" position before making this adjustment.

☐ Connect a DC voltmeter to TP802 (the VCO control line). Set the channel selector to channel 20 and adjust T802 for exactly 2.5 volts at TP802. The voltage on channel 1 should be 1.7 volts and on channel 40, 3.4 volts. Again, make certain that the delta-tune control is in the "O" position.

☐ Connect an rf voltmeter probe to TP803 (the secondary of

T803) and tune T803 for maximum rf indication on channel 20.

□ Connect a frequency counter to TP803 and verify that all frequencies are correct. See Table 8-6 for the correct VCO frequency for each channel. Again, make certain that the delta-tune control is in the "O" position.

□ Connect a frequency counter probe to TP3 to verify that the transmit oscillator is operating at 9.785 MHz. This must be done in the transmit oscillator.

□ Connect an rf voltmeter to TP4 (gate 2 of Q902). In the transmit mode, peak L911 for maximum rf indication on the meter.

Troubleshooting

What do you do if the equipment neither receives nor transmits on all channels? If the PLL circuit is the cause of this problem, the VCO signal at TP803 will either be missing or the signal will be far off frequency and the voltage at pin 22 of IC801 will probably be LOW. If the loop becomes unlocked, the VCO frequency will run wild; that is, out of the control of the loop. However, the VCO should still be operating at some frequency within the 37-MHz vicinity. So if your rf probe shows no indication of rf at TP803, check the VCO oscillator and the VCO buffer amplifier.

If a VCO signal is present at TP803, but it is far off frequency, check the voltage at the lock detector terminal of IC801 at pin 22. If this voltage is low, the loop must be unlocked. Check for the 10.240-MHz reference signal at pin 11 of IC801. If this signal is OK, check at pin 13 (the programmable divider input) for an offset signal. If the offset signal is present, its frequency will not be correct because the VCO frequency is not correct. If the signal is present, this indicates that the control voltage to the VCO control line can't bring the VCO to the correct frequency. The control line voltage at TP802 will probably be abnormally high or abnormally low. There are several possible causes for this. There may be a defective component in the VCO frequency determing circuit, including the varactor diode itself (D802). Maybe the VCO is just out of alignment, usually because of some tinkerer.

If you suspect the VCO to be out of alignment do this: While monitoring the lock detector output voltage at IC801, pin 22, *carefully* and *slowly* tune T802. Try tuning in both directions until you see the voltage at pin 22 go to a HIGH level. If pin 22 goes to a HIGH level, move the DC voltmeter probe to TP802 and adjust T802 for 2.5 volts on channel 20. Needless to say, if tuning T802 doesn't help, make sure you return it to its original position. If the VCO is not at fault, maybe IC802 is defective, but don't replace it unless you have exhausted all other tests. These ICs give little trouble normally. Usually the trouble in the PLL is outside the PLL IC. But don't take this to mean that you will never run across a defective PLL IC—you probably will if you work on enough of these PLL units.

When you checked pin 13 of IC801 for the offset signal, if there was no signal at all there, use an rf voltmeter to trace the signal. If the 11.5966-MHz oscillator is not working, this could cause the offset signal to be missing. Other possible suspects are the VCO buffer (Q806), the loop mixer (Q807), and the driver (Q808).

Incorrect Synthesizer Output Frequency on High Channels with Low Channels OK. This is almost always caused by the VCO being out of alignment. The voltage at TP802 will probably be abnormally high. See the fourth step of the alignment instructions.

Incorrect Synthesizer Output Frequency on Low Channels with High Channels OK. This also is usually caused by VCO misadjustment. The voltage at TP802 will probably be abnormally low. See the fourth step of the alignment instructions.

Wrong Synthesizer Frequency on All Channels. If the loop is locked (as indicated by HIGH level at pin 22 of IC801), the most likely cause is the 11.5966-MHz loop oscillator. Check its frequency with a counter connected to the emitter of Q805 through a small coupling capacitor (5 to 10 pF). There is also a possibility that the reference oscillator may be operating far off frequency. However, it would take a very large change in the 10.240-MHz frequency to cause a small change in the VCO frequency.

No Transmit Frequency Signal. Check the 9.785-MHz oscillator and the transmit mixer stage. Use you rf voltmeter for signal trading.

Wrong Transmit Output Frequency with the VCO OK. If the VCO is OK, then the problem must be the 9.785-MHz oscillator. Use your frequency counter to check the frequency of the 9.785-MHz oscillator.

Some Channels Inoperative. Check for proper programming to pins 1 through 6. Refer to Table 8-6. Check for a defective channel switch and possibly a defective IC801.

Some Channels Intermittent. Check for loose or dirty channel selector contacts. Replace or clean as necessary.

Chapter 9
The MM55106 PLL IC

The MM55106 is a CMOS PLL IC, manufactured by National Semiconductor Corp. The IC is housed in an 18-pin dual in-line plastic package. The outline of the MM55106 is shown in Fig. 9-1. Table 9-1 lists absolute maximum ratings of the MM55106. Table 9-2 lists its various electrical characteristics. Notice that the maximum (guaranteed) toggle frequency is 3 MHz. These devices usually can be operated at much higher input frequencies than the minimum value on the specification sheet. And you will probably find this IC used in circuits where the input to the programmable divider is higher than 3 MHz.

SPECIAL FEATURES

- ☐ Low-power CMOS construction
- ☐ Binary code programming
- ☐ Selectable reference frequency (5 or 10 kHz)
- ☐ 5.12-MHz output
- ☐ On-chip reference oscillator
- ☐ Pulldown resistors on programmable divider inputs
- ☐ Lock detector output

INTERNAL STRUCTURE

Figure 9-2 shows the basic structure of the MM55106. Notice that the 10.240-MHz reference oscillator signal is first divided by 2 to yield a 5.12-MHz signal which when multiplied by 3 can be

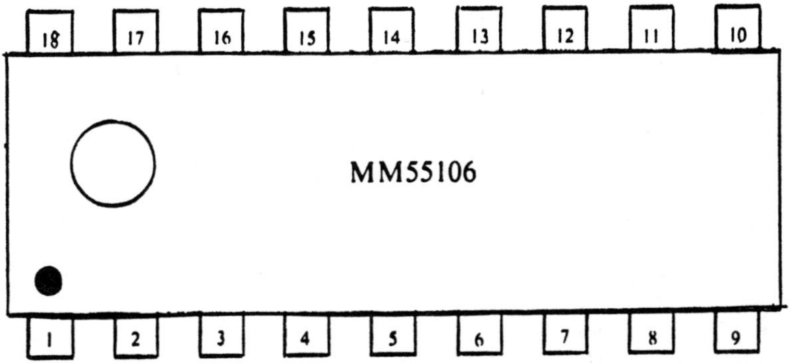

Fig. 9-1. An overview of the MM55106 PLL IC, showing the pin numbering.

Table 9-1. Absolute Maximum Ratings
of the MM55106 (courtesy of National Semiconductor Corp.).

Voltage at any pin ....VCC + 0.3 V to Gnd - 0.3 V

VCC Max....................7 V

Operating temperature.........-30° C to + 75° C

Storage temperature range .....-40° C to + 125° C

Lead temperature (soldering, 10 seconds) 300° C.

mixed with the VCO frequency to provide an offset frequency. This 5.12-MHz signal is further divided by either 5.12 or 1024, depending upon the logic level at the FS (reference frequency select) terminal. If the reference divider is set to divide by 512, this will provide a 10-kHz reference frequency for the phase detector. If the reference divider is set to divide by 1024, this will provide a 5-kHz reference frequency for the phase detector.

Notice in Fig. 9-2 that the programming inputs are labeled P0 through P8 (pins 9 through 17). The weight of the various program pins is shown in parentheses beside the program line. the most significant bit is at pin 9 (P8), which has a weight of 256. The least significant bit is at pin 17 (P0), which has a weight of 1. The

Table 9-2. Electrical Characteristics of the MM55106.

@ TA = 25° C.

PARAMETER	CONDITIONS	MIN	TYP	MAX	UNIT
Supply voltage (VCC)		4.5	5.0	5.5	Volts
Supply current (ICC)	Freq. @ Osc In=10 MHz, @ Fin=2.5 MHz, All other I/0 pins open. VCC = 5.0 V.		3	10	MA.
Logic "1" input voltage (P0 - P8, FS, Fin)		VCC-0.4			V
Logic "0" input voltage (P0 - P8, FS, Fin)				0.4	V
Logic "1" output voltage					
5.12 MHz out, LD	I0 = 0.5 MA				
*o*VCO	I0 = 0.4 MA	VCC-			V
Osc Out	I0 = 0.25 MA	0.5			
Logic "1" input current					
FS-(Pull-up)				1.0	UA
P0 - P8 (Pull-down)	VCC = 5v.	5	20	50	UA
Logic "0" input current					
P0 - P8 (Pull-down)				1.0	UA
FS - (Pull-up)	VCC = 5 v	-10	-35	-100	UA
Maximum Toggle Freq @ Fin.		3			MHz
Maximum Oscillator Freq. @ Osc In		10.24			MHz
TRI-STATE® Leakage @ *o* VCO				1.0	UA

maximum divisor which can be programmed is 511. Each of the programming inputs has an on-chip pulldown resistor tied to it. This simplifies the channel selectors job since it only has to provide binary 1 to the proper pins. Without the pulldown resistors the channel selector would have to ground the unused pins. The pins have the following functions:

□ Pin 1. This pin is the VCC (+5 volts) supply point for the IC. The voltage at this pin should be supplied from a well regulated supply source. The normal VCC voltage should be 5 volts. The minimum VCC voltage should be no less than 4.5 volts and the maximum no more than 5.5 volts.

□ Pin 2. This is the input to the programmable divider. The guaranteed minimum toggle frequency is specified at 3 MHz, though you may find these ICs used at higher frequencies.

□ Pin 3. This is the input to the on-chip reference oscillator.

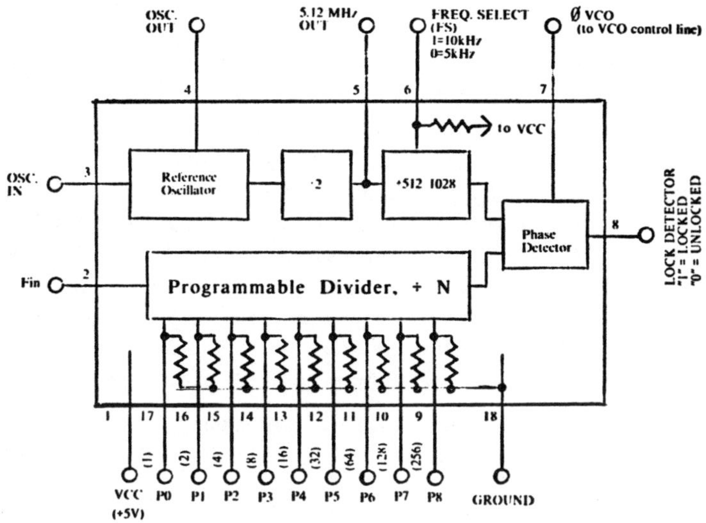

Fig. 9-2. Block diagram of the MM55106. Notice the program inputs (pins 9 through 17) have pulldown resistors while the FS terminal (pin 6) has a pullup resistor (courtesy of National Semiconductor Corp.).

☐ Pin 4. This is the output of the on-chip reference oscillator.

☐ Pin 5. This pin provides a 5.12-MHz signal that can be tripled to 15.360 MHz and then used to mix with the VCO frequency to provide an offset frequency to feed the input of the programmable divider. This helps to minimize the use of crystals.

☐ Pin 6. This is the reference frequency select terminal. There is an on-chip pullup resistor connected to this terminal. If the terminal is left open, the pullup resistor sets the level at binary 1. This sets the reference frequency at 10 kHz. If the terminal is grounded, this sets the level at binary 0, this in turn sets the reference frequency at 5 kHz.

☐ Pin 7. This is the output of the phase detector which controls the VCO frequency. If the VCO frequency is lower than the required lock frequency, this pin will be at a HIGH level. If the VCO frequency is higher than the required lock frequency, this pin will be at a LOW level. If the VCO frequency is equal to the required lock frequency, this pin goes to a high impedance

(TRI-STATE®) condition.

☐ Pin 8. This is the lock detector output. Under normal locked loop conditions, the voltage at pin 8 will be HIGH. However, if the loop fails to achieve lockup, the output from pin 8 will be negative-going pulses. This can be used to trigger a transmit stop circuit.

☐ Pins 9-17. These are the programming inputs to the programmable divider. Each line has an on-chip pulldown resistor connected to it. Programming is achieved by applying binary 1 level to the appropriate pins. To get the N divisor, simply add the weights of the program pins at the binary 1 level. Table 9-3 shows how programming is achieved. The maximum divisor N is 511. This is programmed by applying binary 1 to all program pins (P0 through P8).

☐ Pin 18. This is the ground connection for the IC.

Table 9-3. MM55106 Programming.

N	Weight--	256	128	64	32	16	8	4	2	1
	pin no.---	9	10	11	12	13	14	15	16	17
1		0	0	0	0	0	0	0	0	X
2		0	0	0	0	0	0	0	1	0
.										
.										
.										
.										
511		1	1	1	1	1	1	1	1	1

Notes
Fout = Fin/N
1 = High voltage level
0 = Low voltage level
X = Don't care (either voltage level)

PRACTICAL APPLICATION

Figure 9-3 shows a practical application of the MM55106 PLL IC. This is a little difference arrangement than is found in most PLL circuits; only one crystal is used in this circuit. This is the 10.240-MHz crystal. Therefore, the VCO is required to operate at difference frequencies for the receive and transmit modes. This is accomplished by changing the N divisor when switching from transmit to receive or vice versa.

The programming is done by using two separate wafers of a channel selector switch. Wafer B supplies programming codes for the receive mode, while wafer A supplies programming codes for the transmit mode. Notice that in the receive mode, 5 volts is sup-

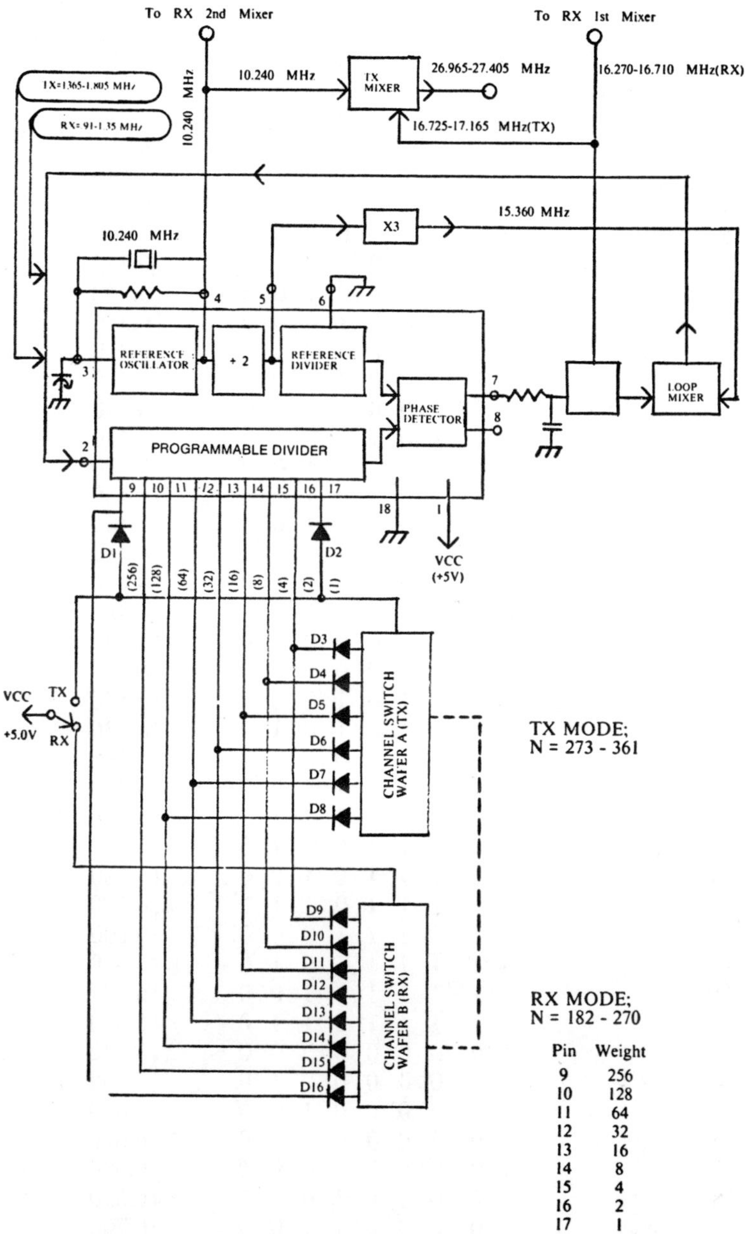

Fig. 9-3. Block diagram of an application of the MM55106 PLL IC. Only one crystal is required for generating receive and transmit frequencies for 40 channels. The program pin weights are listed in the chart in the figure (courtesy of National Semiconductor Corp.).

Table 9-4. Specifications of the PLL Circuit of Fig. 9-3.

CH	N	PRG CD-PN 9-17										VCO Frequency
		Weight	256	128	64	32	16	8	4	2	1	MHz
		PIN-------	9	10	11	12	13	14	15	16	17	
1	182		0	1	0	1	1	0	1	1	0	16.270
2	184		0	1	0	1	1	1	0	0	0	16.280
3	186		0	1	0	1	1	1	0	1	0	16.290
4	190		0	1	0	1	1	1	1	1	0	16.310
5	192		0	1	1	0	0	0	0	0	0	16.320
6	194		0	1	1	0	0	0	0	1	0	16.330
7	196		0	1	1	0	0	0	1	0	0	16.340
8	200		0	1	1	0	0	1	0	0	0	16.360
9	202		0	1	1	0	0	1	0	1	0	16.370
10	204		0	1	1	0	0	1	1	0	0	16.380
11	206		0	1	1	0	0	1	1	1	0	16.390
12	210		0	1	1	0	1	0	0	1	0	16.410
13	212		0	1	1	0	1	0	1	0	0	16.420
14	214		0	1	1	0	1	0	1	1	0	16.430
15	216		0	1	1	0	1	1	1	0	0	16.440
16	220		0	1	1	0	1	1	1	0	0	16.460
17	222		0	1	1	0	1	1	1	1	0	16.470
18	224		0	1	1	1	0	0	0	0	0	16.480
19	226		0	1	1	1	0	0	0	1	0	16.490
20	230		0	1	1	1	0	0	1	1	0	16.510
21	232		0	1	1	1	0	1	0	0	0	16.520
22	234		0	1	1	1	0	1	0	1	0	16.530
23	240		0	1	1	1	1	0	0	0	0	16.560
24	236		0	1	1	1	0	1	1	0	0	16.540
25	238		0	1	1	1	0	1	1	1	0	16.550
26	242		0	1	1	1	1	0	0	1	0	16.570
27	244		0	1	1	1	1	0	1	0	0	16.580
28	246		0	1	1	1	1	0	1	1	0	16.590
29	248		0	1	1	1	1	1	0	0	0	16.600
30	250		0	1	1	1	1	1	0	1	0	16.610
31	252		0	1	1	1	1	1	1	0	0	16.620
32	254		0	1	1	1	1	1	1	1	0	16.630
33	256		1	0	0	0	0	0	0	0	0	16.640
34	258		1	0	0	0	0	0	0	1	0	16.650
35	260		1	0	0	0	0	0	1	0	0	16.660
36	262		1	0	0	0	0	0	1	1	0	16.670
37	264		1	0	0	0	0	1	0	0	0	16.680
38	266		1	0	0	0	0	1	0	1	0	16.690
39	268		1	0	0	0	0	1	1	0	0	16.700
40	270		1	0	0	0	0	1	1	1	0	16.710

Receive Mode

Table 9-5. Specifications of the PLL Circuit of Fig. 9-3.

CN	N	PRG CD-PN 9-17									VCO
		Weight 256	128	64	32	16	8	4	2	1	FREQ
		PIN-------9	10	11	12	13	14	15	16	17	MHz
1	273	1	0	0	0	1	0	0	0	1	16.725
2	275	1	0	0	0	1	0	0	1	1	16.735
3	277	1	0	0	0	1	0	1	0	1	16.765
4	281	1	0	0	0	1	1	0	0	1	16.765
5	283	1	0	0	0	1	1	0	1	1	16.775
6	285	1	0	0	0	1	1	1	0	1	16.785
7	287	1	0	0	0	1	1	1	1	1	16.795
8	291	1	0	0	1	0	0	0	1	1	16.815
9	293	1	0	0	1	0	0	1	0	1	16.825
10	295	1	0	0	1	0	0	1	1	1	16.835
11	297	1	0	0	1	0	1	0	0	1	16.845
12	301	1	0	0	1	0	1	1	0	1	16.865
13	303	1	0	0	1	0	1	1	1	1	16.875
14	305	1	0	0	1	1	0	0	0	1	16.885
15	307	1	0	0	1	1	0	0	1	1	16.895
16	311	1	0	0	1	1	0	1	1	1	16.915
17	313	1	0	0	1	1	1	0	0	1	16.925
18	315	1	0	0	1	1	1	0	1	1	16.935
19	317	1	0	0	1	1	1	1	0	1	16.945
20	321	1	0	1	0	0	0	0	0	1	16.965
21	323	1	0	1	0	0	0	0	1	1	16.975
22	325	1	0	1	0	0	0	1	0	1	16.985
23	331	1	0	1	0	0	1	0	1	1	17.015
24	327	1	0	1	0	0	0	1	1	1	16.995
25	329	1	0	1	0	0	1	0	0	1	17.005
26	333	1	0	1	0	0	1	1	0	1	17.025
27	335	1	0	1	0	0	1	1	1	1	17.035
28	337	1	0	1	0	1	0	0	0	1	17.045
29	339	1	0	1	0	1	0	0	1	1	17.055
30	341	1	0	1	0	1	0	1	0	1	17.065
31	343	1	0	1	0	1	0	1	1	1	17.075
32	345	1	0	1	0	1	1	0	0	1	17.085
33	347	1	0	1	0	1	1	0	1	1	17.095
34	349	1	0	1	0	1	1	1	0	1	17.105
35	351	1	0	1	0	1	1	1	1	1	17.115
36	353	1	0	1	1	0	0	0	0	1	17.125
37	355	1	0	1	1	0	0	0	1	1	17.135
38	357	1	0	1	1	0	0	1	0	1	17.145
39	359	1	0	1	1	0	0	1	1	1	17.155
40	361	1	0	1	1	0	1	0	0	1	17.165

| Transmit Mode

plied to wafer B through the transmit/receive switch. In the receive mode, wafer B applies 5 volts (binary 1) in various combinations to programming pins 9 through 16. The receive mode N value ranges from 182 for channel 1 to 270 for channel 40. Table 9-4 lists the receive mode N number, programming code, and VCO frequency for each channel for the PLL circuit of Fig. 9-3. Notice that the N numbers increase in increments of 2 when 10-kHz channel spacing exists. This is because the reference frequency for the phase detector is 5 kHz, as determined by the grounding of pin 6. Table 9-4 shows that the binary level at pin 17 is always 0 in the receive mode. For this reason, pin 17 is left open in the receive mode. All the other program pins connect to wafer B on the channel selector switch.

In the transmit mode, the transmit/receive switch removes the +5 volts from the common terminal of wafer B and applies the +5 volts to the common terminal of wafer A. In the tramsit mode, wafer B applies the binary 1 (5 volts) in various combinations to program pins 10 through 16. The transmit mode N value ranges from 273 to 361. Table 9-5 lists the N value, programming code, and VCO frequency for each channel. Notice that pins 9 and 17 are always at the binary 1 level. Thus, they can be connected directly to +5 volts. Figure 9-3 shows that pins 9 and 17 are connected to +5 volts through diodes D1 and D2, as long as the transmit/receive switch is in the transmit mode. Diodes D1 and D2 provide isolation between pins 9 and 17 in the receive mode. Also, Table 9-5 shows that program pin 10 is always at binary 0 in the transmit mode. For this reason, pin 10 is left open in the transmit mode. In the transmit mode, only pins 11 through 16 connect to the channel switch. Diodes D3 through D16 provide isolation between the two wafers of the channel switch so that there is no interaction between the two wafers.

Let's analyze the PLL circuit action for CB channel 1. First, we'll cover the receive mode. When the channel selector is set for channel 1 and the set operating in the receiving mode, an N value of 182 is programmed into the programmable divider. The output frequency of the programmable divider must be 5 kHz to achieve lock. Thus, the input to the programmable divider would have to be 5 kHz × 182, or 0.91 MHz. As you can see from Figure 9-3, the 5.12-MHz signal at pin 5 of the MM55106 is multiplied by 3 to yield 15.360 MHz. This 15.360-MHz signal is then applied to the loop mixer. The VCO signal is also fed to the loop mixer. The offset frequency appearing at pin 2 (programmable divider input) is equal to the VCO frequency minus the 15.360-MHz frequency. The phase detector output forces the VCO to operate at a frequency which is higher than 15.60 MHz by an amount equal to n × 5 kHz (the reference frequency). Thus for channel 1, the receive mode VCO frequency would be fVCO (MHz) = 15.360 + 0.005N. It would be fVCO = 15.360 + 0.005(182) = 15.360 + 0.91, or 16.270 MHz. A portion of this 16.270-MHz VCO signal is fed to the receive first mixer where it is heterodyned with the incoming receive signals. The output of the first mixer usually contains a ceramic filter which passes 10.695 MHz and rejects other frequencies. When the 16.270-MHz VCO signal is beat against a 26.965-MHz channel 1 signal, a 10.695-MHz difference frequency appears at the output of the mixer. This is passed by the 10.695-MHz ceramic filter on to the second mixer. In the second mixer, the 10.695-MHz first i-f signal is heterodyned with a 10.240-MHz signal from the reference oscillator. The difference of these two frequencies is 455 kHz. A 455-kHz ceramic filter in the output of the second mixer passes this 455-kHz second intermediate frequency. This 455-kHz signal is then amplified and detected.

Now let's look at the channel 1 transmit mode of operation. When the push-to-talk switch on the microphone is pressed, the +5 volts (VCC) is removed from wafer B of the channel switch and applied to wafer A. The programming is then done by wafer A. In the transmit mode, pins 9 and 17 are connected to VCC through diodes D1 and D2. Thus, the combined weight of pins 9 and 17 will always be used in determining the N number. The combined weight of pins 9 and 17 is 257 (256 + 1). Also, in the transmit mode, pin 10 is left open. The on-chip pulldown resistor automatically places pin 10 at ground or binary 0. The other six pins (11 through 16) connect to the channel switch. Wafer A applies +5 volts (binary 1) to the appropriate program pins to achieve the proper N number. The proper N number for the channel 1 transmit mode is shown in Table 9-5. This table also shows the proper program code and VCO frequency for each channel in the transmit mode. You can see from the table that the N number for channel 1 transmit should be 273. Since the output of the programmable divider has to be 5 kHz to achieve lock, the input to the programmable divider will have to be 5 kHz × 273, or 1.365 MHz. It was previously established that the VCO frequency for this PLL could be determined by the formula, fVCO = 15.360 + 0.005N. Substituting, you have fVCO = 15.360 + 0.005(273) = 15.360 + 1.365, or 16.725 MHz. This 16.725-MHz VCO signal is fed to a transmit

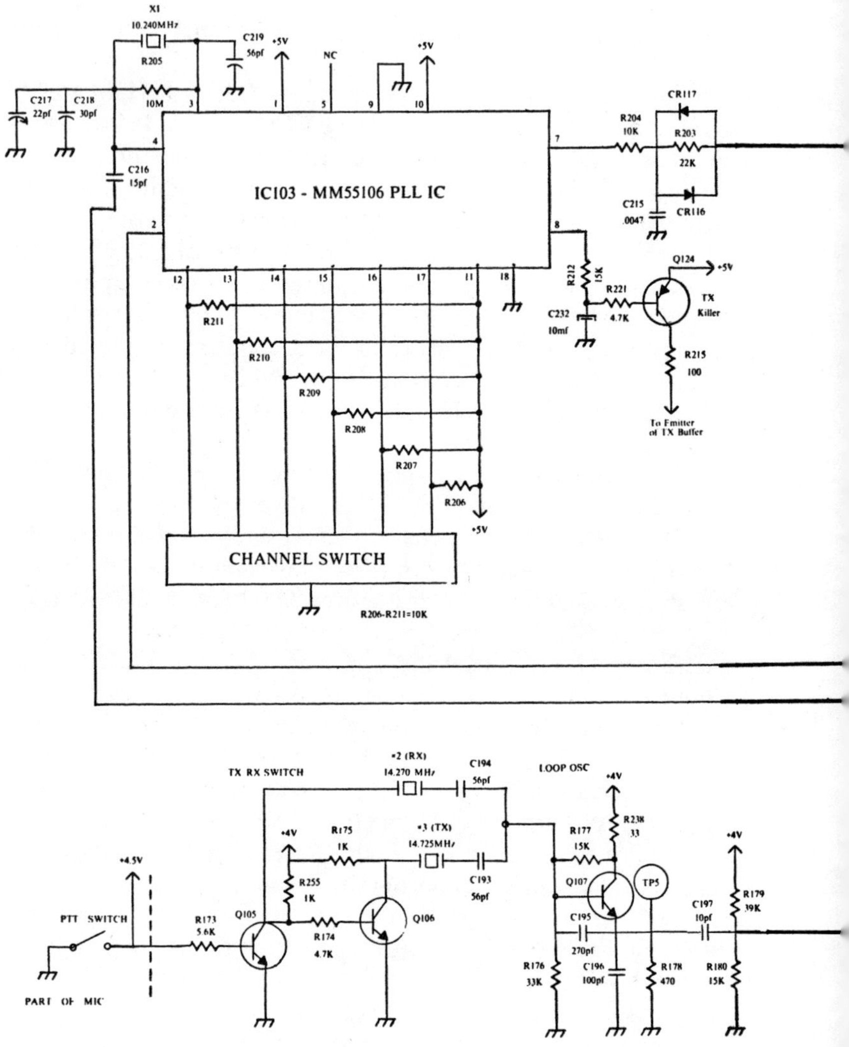

Fig. 9-4. Schematic of the PLL circuit used in the Regency model CB-501, 40-channel CB radio.

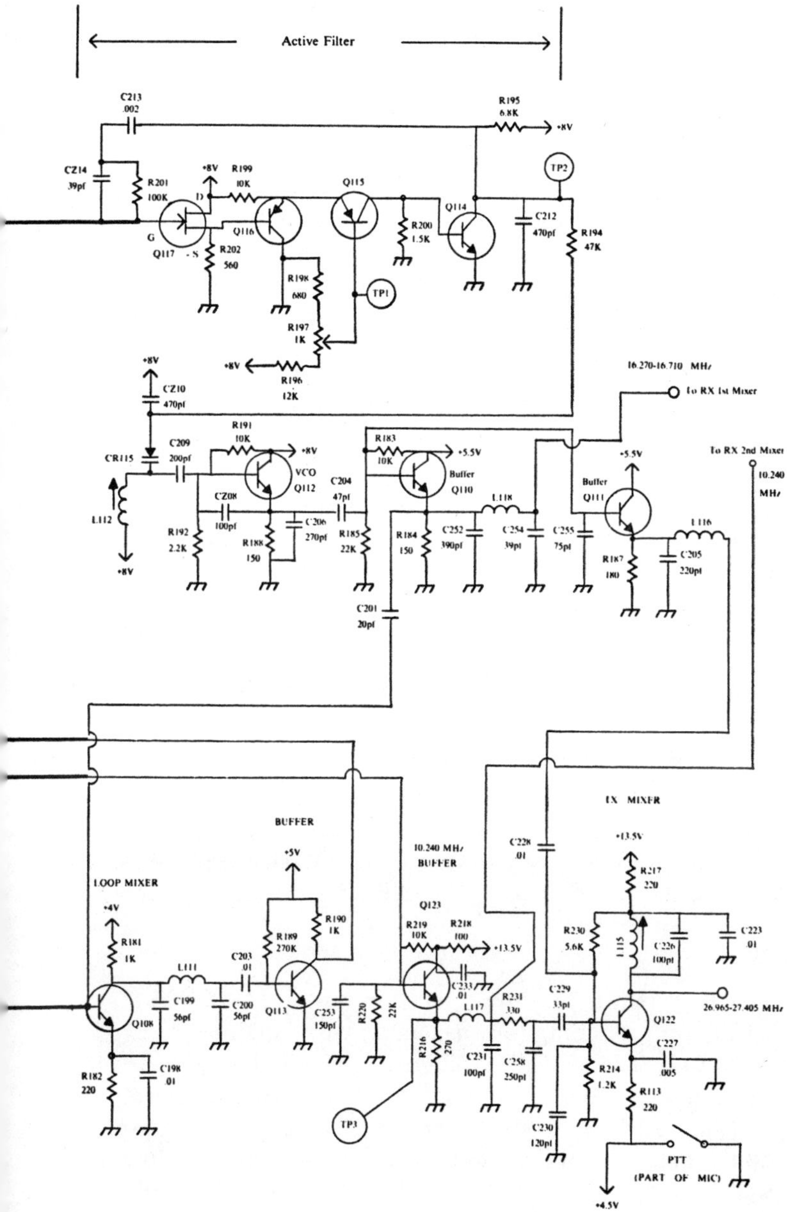
Active Filter
C213
.002
R195
6.8K
+8V
TP2
CZ14
39pf
R201
100K
+8V
R199
10K
Q115
Q114
C212
470pf
R194
47K
R200
1.5K
Q116
G
Q117
R202
560
R198
680
TP1
R197
1K
R196
12K
+8V
16.270-16.710 MHz
To RX 1st Mixer
CZ10
470pf
R191
10K
C209
200pf
CR115
VCO
Q112
R183
10K
+5.5V
Buffer
Q110
+5.5V
Buffer
Q111
To RX 2nd Mixer
10.240
MHz
L112
C204
47pf
CZ08
100pf
C206
270pf
L118
L116
R192
2.2K
R188
150
R185
22K
R184
150
C252
390pf
C254
39pf
C255
75pf
R187
180
C205
220pf
C201
20pf
TX MIXER
BUFFER
+5V
10.240 MHz
BUFFER
C228
.01
+13.5V
R217
220
LOOP MIXER
+4V
R181
1K
R189
270K
R190
1K
Q123
R219
10K
R218
100
+13.5V
R230
5.6K
L115
C226
100pf
C223
.01
L111
C203
.01
C199
56pf
C200
56pf
Q113
Q108
C253
150pf
R220
22K
C233
.01
L117
R231
330
C229
33pf
26.965-27.405 MHz
Q122
C227
.005
R216
270
C231
100pf
C258
250pf
R214
1.2K
R113
220
R182
220
C198
.01
TP3
C230
120pf
PTT
(PART OF MIC)
+4.5V

mixer where it is mixed with a 10.240-MHz signal from the reference oscillator. The sum of these two frequencies is 16.725 MHz + 10.240 MHz, or 26.965 MHz, which is the correct channel 1 transmit frequency. This signal is then amplified by the following transmit stages.

This particular PLL circuit does have the advantage of using only 1 crystal to generate all the required transmit and receive frequencies. However, the channel selector switch is very complex, requiring two wafers. Also, wafer B has to supply programming to eight program lines. This makes it more complex. The majority of PLL circuits only require channel switching to six program lines. Further complexity is caused by using a 5-kHz reference frequency. This requires a range of N numbers that is twice that required when a 10-kHz reference is used.

Actuall, it might have been more economical to have used another crystal or two rather than such a complex channel switch arrangement. This particular application is presented not because it was so practical but because it shows that there are different ways of achieving the same result. The method used usually depends on the manufacturer's own preference, the availability of parts, and certainly the cost of the parts. This circuit was based on one presented in National Semiconductor's data sheet on the MM55106.

THE REGENCY CB-501 PLL CIRCUIT

The schematic diagram of the Regency CB-501 PLL circuit is shown in Fig. 9-4. This set uses the MM55106 as the heart of the PLL digital frequency synthesizer circuit. The function and operation of the various stages of this PLL circuit are as follows:

PLL IC

You may recall from the discussion on the MM55106 that each of the programming pins has an on-chip pulldown resistor connected to it. This pulls the voltage on that pin to a LOW level (binary 0) if the pin is left open. Normally, a channel selector switch would apply +5 volts (binary 1) to the appropriate pins, leaving the pins open which are to be at binary 0. In this circuit, however, external pullup resistors have been installed that are capable of overriding the on-chip pulldown resistors. Only pins 12 through 17 connect to the channel selector switch so only these pins have the external pullup resistors connected to them.

With this arrangement, the channel selector grounds the pins which are supposed to be at binary 0. Then the pins which are supposed to be at binary 1 are pulled up to the binary 1 level by the pullup resistors. Notice that program pin 9, with a weight of 256, is grounded. This means that the weight of 256 will never be used in determining the N value. Also, pins 10 and 11 are connected to +5 volts (binary 1), so there respective weights of 128 and 64 will always be used in determining the N values for each of the 40 channels. This means that the combined weight of pins 10 and 11, which is 192 will be added to the number programmed into pins 12 through 17 to get the total N value. Table 9-6 lists the N value, program code and VCO frequencies for each channel. Notice that on channel 1, the number programmed into pins 12 through 17 is 8 because pin 14, with a weight of 8, is the only one of these pins at the binary 1 level. To get the total N value, simply add 192 + 8 = 200.

Notice that pin 6, which is the reference frequency selector pin, is left open. The on-chip pullup resistor places pin 6 at 5 volts or binary 1. And a binary 1 at pin 6 selects a reference frequency of 10 kHz.

The reference oscillator utilizes the on-chip oscillator by connecting a feedback resistor and a 10.240-MHz crystal between pins 3 and 4 of IC103. The frequency of the reference oscillator can be finely tuned with C217.

Voltage-Controlled Oscillator (VCO) and Buffer

The voltage-controlled oscillator is built around transistor Q112. Capacitor C208, which connects between the emitter and base, provides feedback for oscillation. The resonant circuit of the VCO consists of L112, C210, and varactor diode CR115. The VCO operates in the 16 to 17-MHz range. Coil L112 can be considered a *course frequency adjustment* while the control voltage at TP2 serves as a *fine frequency adjustment*. This control voltage varies the reverse bias on the varactor through R194 to fine tune the VCO to a frequency that will enable the loop to lock and remain locked. Notice that the control voltage is applied to the anode of the varactor.

The VCO signal is fed from the emitter of Q112 through C204 to the base of buffer Q110 and also to the base of buffer Q111. Buffer Q110 feeds the VCO signal from its emitter through a low-pass filter on to the receive first mixer. Buffer Q111 feeds the VCO signal from its emitter through a low-pass filter on to the transmit mixer stage. Buffer transistors Q110 and Q111 are connected as emitter followers.

Loop Oscillator

The loop oscillator is built around Q107. Capacitor C195, which connects between the emitter and base of Q107, provides feedback for oscillation. Notice that there are two crystals associated with the oscillator. One is for the receive mode; the other is for the transmit mode. In the receive mode, transistor Q105 receives bias through R173. This causes Q105 to saturate, thus grounding one end of crystal X2, which is resonant at 14.270 MHz. When Q105 is turned on (in the receive mode), this kills the bias to transistor Q106. Thus, when Q105 is on, Q106 is off. In the receive mode, crystal X2 is grounded while crystal X3 is floating or disconnected from ground. Therefore, in the receive mode, crystal X2 sets the oscillator frequency at 14.270 MHz.

In the transmit mode, when the PTT switch is closed, the forward bias to Q105 is killed. Thus, Q105 turns off. With Q105 turned off, crystal X2 is ungrounded. Also, when Q105 is turned off, it no longer kills the forward bias to Q106, so Q106 turns on grounding crystal X3. This sets the oscillator frequency at 14.725 MHz.

The oscillator signal is fed from the emitter of Q107 to the loop mixer through C197. Notice that the emitter of Q107 is TP5 (test point 5).

Loop Mixer

The loop mixer is a simple common-emitter circuit whose base is fed two input signals, one from the 14-MHz loop oscillator and the other from the VCO buffer, Q110. The VCO frequency is mixed with the loop oscillator frequency to provide a difference frequency at the mixer output. This difference frequency ranges from 2.00 MHz to 2.44 MHz. This difference frequency is fed through a low-pass filter on to the buffer transistor, Q113. this buffer amplifies the difference signal and feeds it on to the input of the programmable divider at pin 2 of IC103.

10.240-MHz Buffer

This stage receives its input from the 10.240-MHz reference oscillator and feeds the 10.240-MHz signal to the transmit mixer and to the second receive mixer. The emitter of the 10.240-MHz buffer is designated as TP3. This buffer stage is operated as an emitter follower. Notice the output signal is taken from the emitter. the collector is placed at rf ground by C233, a 0.01-muF capacitor.

Transmit Mixer

This transistor mixes the VCO signal with the 10.240-MHz reference signal to produce a sum frequency that is equal to the transmit frequency. In the receive mode, the mixer transistor is disabled because the emitter loses its ground through the PTT switch.

Transmit Killer

Transistor Q124 receives its input from the lock detector output of the MM55106. During normal locked loop conditions, the voltage at pin 8 (LD) is 5 volts. This voltage appears at the base of Q124. Since the voltage at the emitter of Q124 is also at +5 volts, Q124 will be turned off. However, if the loop loses lock, the voltage at the base of Q124 will then drop, causing Q124 to become forward biased. When Q124 turns on, the collector voltage rises to near +5 volts. The +5 volts is applied to the emitter of the transmit buffer (not shown). This reverse biases the transmit buffer stage, thus disabling the transmitter.

Active Filter

The active filter consists of four transistor stages: Q117, Q116, Q115, and Q114. Notice that direct coupling is used throughout the four stages so that a change in the DC voltage level at the gate of Q117 affects the DC level at TP2, the collector of Q114. Small voltage changes at the input are converted into large voltage changes at TP2. Furthermore, the voltage change at TP2 will be in an opposite direction to the voltage change at the gate of Q117. If the positive voltage at the gate of Q117 should increase, the voltage at TP2 will decrease. A decrease in the voltage at the gate of Q117 will cause an increase in the voltage at TP2. This is why the varactor in the VCO has its anode connected to the control line and its cathode connected to a fixed positive voltage through L112.

If the VCO frequency is higher than the required lock frequency, the output of the phase detector at pin 7 of IC103 will drop to a LOW level. This drop in voltage at the gate of Q117 causes the voltage at TP2 to increase. This increased voltage on the VCO control line reduces the reverse bias on the varactor, CR115. This increases the capacitance of the varactor, causing the VCO frequency to decrease. If the VCO frequency is low, just the opposite situation exists. Thus, the VCO is steered to just the right frequency.

LED Channel Display Circuit

Figure 9-5 shows a diagram of the LED channel display circuit. This is a common-anode type of display. All anodes connect to pins 13 and 14. All of the cathodes are connected to the channel selector switch through individual current limiting resistors. The channel switch ground the appropriate segments. Notice that in the "brite" position, the common terminals (13 and 14) are connected directly to the +5-volts supply. In the "dim" position, three diodes are connected in series with the LED. This drops the voltage to reduce the brightness of the LED segments. The rear view of the LED is shown in Fig. 9-6.

Analysis of Operation

When the channel selector is set for channel 40, the N value

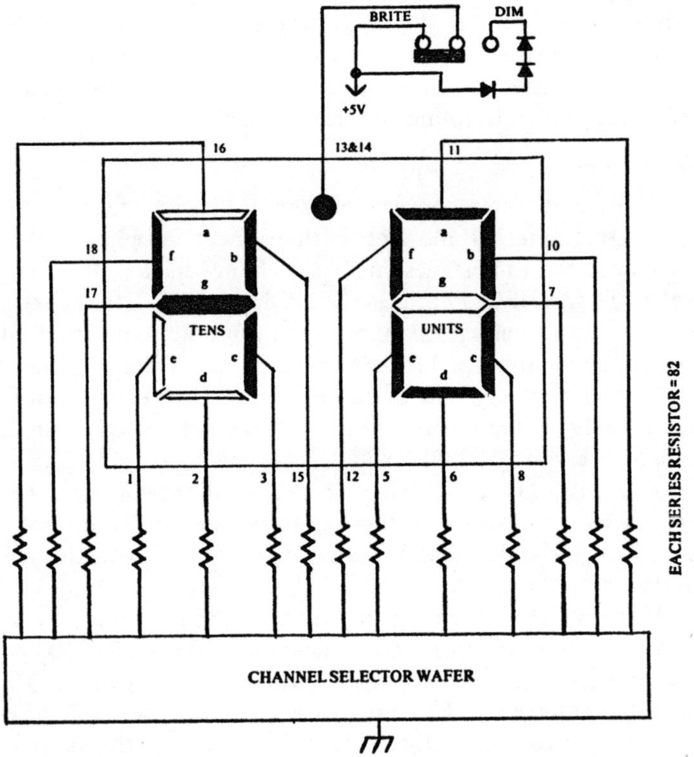

Fig. 9-5. Drawing of the LED display used in the Regency CB-501. This is a common-cathode type of LED display. The darkened segments represent channel 40.

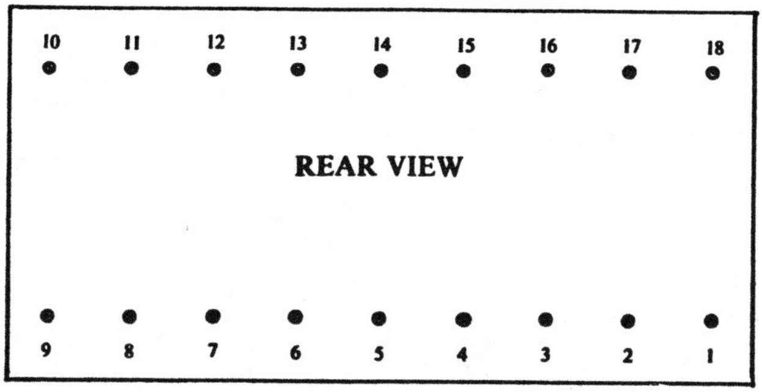

Fig. 9-6. A rear view of the LED display that shows the pin numbering.

programmed into the programmable divider is 244. Table 9-6 lists the various N values, programming codes and VCO frequencies for the 40 channels. From Table 9-6, you can see that the number programmed into pins 12 through 17 is 32 + 16 + 4, or 52. To get the total N value, add 192 to this to get 192 + 52, or 244.

From Table 9-6, the correct VCO frequency for the channel 40 receive mode is 16.710 MHz. When the set is first turned on, the VCO will start operating in the 17-MHz vicinity. For example, suppose that the VCO started operating at 17.000 MHz exactly. This VCO signal will be fed from the emitter of VCO buffer Q110 to the base of the loop mixer, Q108. Also, a 14.270-MHz signal from the loop oscillator feeds the base of the loop mixer, Q108. The loop mixer mixes these two frequencies, and the collector circuit of Q108 passes the difference frequency while attenuating the other frequencies. This difference frequency will be 17.000 MHz - 14.270 MHz, or 2.73 MHz. This 2.73-MHz signal is amplified by buffer Q113 and fed to the input of the programmable divider at pin 2 of IC103. The programmable divider divides this 2.73 MHz frequency by 244 to yield a frequency of 11.1885 kHz at the output of the programmable divider. Because this frequency is higher than the 10-kHz phase detector reference frequency, the output voltage from the phase detector at pin 7 will drop. this voltage drop at pin 7 of IC103 will cause the voltage at TP2 (VCO control line) to increase. This increased control line voltage decreases the VCO frequency. As the VCO frequency decreases, the input to the programmable divider approaches 2.44 MHz, the output from the programmable divider approaches 10 kHz, and the error signal from the phase detector grows smaller. When the

Table 9-6. Specifications of the Regency CB-501 PLL Circuit.

CH.	N	Program code to pins 9-17 Weight---- 32 16 8 4 2 1 Pin No.---- 12 13 14 15 16 17	TX MD VCO Frq.-MHz	RX MD VCO Frq.-MHz
1	200	0 0 1 0 0 0	16.725	16.270
2	201	0 0 1 0 0 1	16.735	16.280
3	202	0 0 1 0 1 0	16.745	16.290
4	204	0 0 1 1 0 0	16.765	16.310
5	205	0 0 1 1 0 1	16.775	16.320
6	206	0 0 1 1 1 0	16.785	16.330
7	207	0 0 1 1 1 1	16.795	16.340
8	209	0 1 0 0 0 1	16.815	16.360
9	210	0 1 0 0 1 0	16.825	16.370
10	211	0 1 0 0 1 1	16.835	16.380
11	212	0 1 0 1 0 0	16.845	16.390
12	214	0 1 0 1 1 0	16.865	16.410
13	215	0 1 0 1 1 1	16.875	16.420
14	216	0 1 1 0 0 0	16.885	16.430
15	217	0 1 1 0 0 1	16.895	16.440
16	219	0 1 1 0 1 1	16.915	16.460
17	220	0 1 1 1 0 0	16.925	16.470
18	221	0 1 1 1 0 1	16.935	16.480
19	222	0 1 1 1 1 0	16.945	16.490
20	224	1 0 0 0 0 0	16.965	16.510
21	225	1 0 0 0 0 1	16.975	16.520
22	226	1 0 0 0 1 0	16.985	16.530
23	229	1 0 0 1 0 1	17.015	16.560
24	227	1 0 0 0 1 1	16.995	16.540
25	228	1 0 0 1 0 0	17.005	16.550
26	230	1 0 0 1 1 0	17.025	16.570
27	231	1 0 0 1 1 1	17.035	16.580
28	232	1 0 1 0 0 0	17.045	16.590
29	233	1 0 1 0 0 1	17.055	16.600
30	234	1 0 1 0 1 0	17.065	16.610
31	235	1 0 1 0 1 1	17.075	16.620
32	236	1 0 1 1 0 0	17.085	16.630
33	237	1 0 1 1 0 1	17.095	16.640
34	238	1 0 1 1 1 0	17.105	16.650
35	239	1 0 1 1 1 1	17.115	16.660
36	240	1 1 0 0 0 0	17.125	16.670
37	241	1 1 0 0 0 1	17.135	16.680
38	242	1 1 0 0 1 0	17.145	16.690
39	243	1 1 0 0 1 1	17.155	16.700
40	244	1 1 0 1 0 0	17.165	16.710

Note: To get the N value add 192 to the number programmed into pins 12-17.

VCO frequency reaches 16.710 MHz, the input to the programmable divider will be 2.44 MHz and the output from the programmable divider will be 10 kHz. At this point, the loop will lock and the output of the phase detector will contain no error signal component. Thus, the VCO frequency stabilizes at 16.710 MHz. This 16.710-MHz signal is fed to the receive first mixer, where it is mixer with a 27.405-MHz, channel 40 signal to produce a difference frequency of 10.695 MHz. This is the first intermediate frequency. This is fed on to a second mixer, where it is heterodyned with a signal from the 10.240-MHz reference oscillator buffer to produce a difference frequency of 455 kHz (10.695 MHz - 10.240 MHz). This is the second intermediate frequency. This 455-kHz i-f is then amplified and detected.

What happens in the transmit mode? When the push-to-talk switch is pressed, transistor Q105 is turned off and Q106 is turned on. This disconnects crystal X2 from the circuit and connects X3 in the circuit. The oscillator now operates at 14.725 MHz. At the instant that the transmitter is keyed, the VCO will still be operating at 16.710 MHz. This is mixed with the 14.725-MHz signal from the loop oscillator to produce a difference frequency of 16.710 MHz - 14.725 MHz, or 1.985 MHz at the output of the loop mixer. This 1.985-MHz signal is fed to buffer Q113 where it is amplified and fed to the input of the programmable divider at pin 2 of IC103. The programmable divider divides this 1.985-MHz signal by 244 to yield a frequency of 8.135 kHz at the output of the programmable divider. Since this is lower than the 10-kHz phase detector reference frequency, the phase detector issues a correction signal which causes the voltage at pin 7 of IC103 to increase. This voltage increase at pin 7 causes the VCO control voltage at TP2 to decrease. This increases the VCO frequency. As the VCO frequency approaches 17.165 MHz the output of the programmable divider approaches 10 kHz and the phase detector error signal decreases. When the VCO frequency reaches 17.165 MHz, the input to the programmable divider will be 2.44 MHz, the output from the programmable divider will be 10 kHz, and the control voltage from IC103 pin 7 will no longer contain an error component. The VCO frequency stabilizes at 17.165 MHz and the loop locks.

In the transmit mode, the VCO signal is fed from Q111 to the base of the transmit mixer, Q122. Also, a 10.240-MHz signal from buffer Q123 feeds the base of the transmit mixer. The 17.165-MHz VCO signal is mixed with the 10.240-MHz reference signal to produce a sum frequency of 27.405 MHz. This is the correct channel 40 transmit frequency. The output of the mixer, Q122, is tuned to this frequency. All the other frequencies are attenuated. Notice that the emitter of Q122 is grounded through resistor R213 and the PTT switch contacts. In the receive mode, this PTT switch is open, thus disabling the Tx mixer.

Alignment

☐ Connect a DC voltmeter probe to TP1 (the base of Q115) and adjust R197 for exactly + 0.75 volts. This can be done on any channel.

☐ Connect a frequency counter to TP3 (the emitter of Q123) and adjust C217 for 10.240 MHz + 100 Hz. This can be done on any channel.

☐ Set the channel selector to channel 19 and connect a DC voltmeter probe to TP2 (the collector of Q114). Adjust L112 for exactly 2.0 volts at TP2.

☐ Connect a frequency counter to TP5 (the emitter of Q107) and in the receive mode, check for 14.270 MHz. The frequency here should be within + 500 Hz.

☐ With the frequency counter still connected to TP5, key the transmitter and check for 14.725 MHz. This frequency should be within + 250 Hz.

☐ Connect a frequency counter to the emitter of Q110 and check for the proper VCO frequencies in both the transmit receive modes. See Table 9-6 for the correct VCO frequencies for each channel. the receive mode VCO frequencies should be within +500 Hz. The transmit mode VCO frequencies should be within + 250 Hz.

Troubleshooting

If the rig neither transmits nor receives on any channel, first check the supply voltages to the PLL circuit. If the supply voltages are correct, check for rf voltage at the emitter of the VCO buffer, Q110, and also check the frequency of the signal at this point. If the PLL circuit is at fault, the VCO signal will probably be missing or will be operating far from the correct frequency. If no rf voltage is measured at the emitter of Q110, troubleshoot the VCO (Q112) and the VCO buffer (Q110).

If you do find rf voltage at the emitter of Q110, but the frequency is far from correct, check the voltage level at the lock detector terminal, pin 8 of IC103. If the voltage there is low, the loop is unlocked. This can be caused by many different things. Use an rf voltmeter to check for the presence of the 10.240 MHz oscillator signal at pin 4 of IC103. Also, use a frequency counter to verify that the 10.240-MHz signal is correct. If this signal is missing or the frequency is incorrect, troubleshoot the 10.240-MHz oscillator. Since this is an on-chip oscillator, all you can do is check the external components, such as the crystal, capacitors, and resistor R205. If the on-chip oscillator itself is defective, the IC will have to be replaced. If the 10.240-MHz reference signal is OK, check for rf voltage at pin 2 of IC103. If there is a good rf voltage indication there, this indicates that the loop oscillator, loop mixer, and loop buffer stages are working. In this case the VCO oscillator is the prime suspect. Check all the stages of the active filter to make sure that no defect there is causing the VCO to run off frequency. Check all components in the frequency-determining circuit of the VCO oscillator. This includes the varactor, CR115. If no defects are found, it may be that the VCO oscillator has been mistuned. In this case you can get it realigned by the following method. Monitor the DC voltage level at pin 8 of IC103 while tuning L112. Tune L112 until the voltage at pin 8 goes HIGH, indicating a locked loop. Then monitor the voltage at TP2 and tune L112 for exactly 2 volts on channel 19. If you get no response with this procedure, return L112 to its original position. When all else fails, you might have to resort to a substituting IC103, but only as a last resort.

If there is no indication of rf voltage at pin 2 of IC103, use an rf voltmeter to signal trace through the loop buffer, loop mixer, and loop oscillator.

High Channels Inoperative with Low Channels OK. This is usually caused by improper alignment of the VCO. The voltage at TP2 will probably be abnormally low if misalignment of the VCO is causing this trouble. See the third step of the alignment instructions to correct this. If this is not the trouble, check the active filter stages for any defect.

Low Channels Inoperative with High Channels OK. This is usually caused by improper alignment of the VCO. The voltage at TP2 will probably be abnormally high if misalignment of the VCO is the trouble. See the third step 3 of the alignment instructions to correct this. If this is not the trouble, check the active filter stages.

Wrong Transmit Frequencies on All Channels. Troubleshoot the 14-MHz loop oscillator. Also check the crystal switching circuit also. Check the frequency at TP5. It should be 14.725 MHz.

Wrong Receive Frequencies on All Channels. Troubleshoot the 14-MHz loop oscillator. Check the crystal switching circuit. Check the frequency at TP5. It should be 14.270 MHz.

No Transmit Frequency Signal with Receive OK. Check the 14-MHz oscillator. Check buffer Q111 and transmit mixer Q122.

No Receive with Transmit OK. Check the 14-MHz oscillator. Check for 10.240 MHz at TP3. If it is missing, check buffer Q123.

Some Channels Inoperative. Check the programming to IC103. Check the channel switch and check for open or shorted program lines. Possibly, a defective IC103 is the cause.

Some Channels Intermittent. This is caused by loose or dirty channel switch contacts. Clean or replace as necessary.

Chapter 10 The MN6040 PLL IC Series

The MN6040 series of PLL IC devices is manufactured by Matsushita Electronics Corporation (Panasonic). There are three separate PLL ICs in the series: MN6040, MN6040A, and MN60-40B. Each of these devices are very similar, yet important differences exist between them. Each of these devices is a CMOS IC especially designed for PLL synthesizer circuits. Each device contains a reference frequency divider, programmable divider, and phase detector. The MN6040A and the MN6040 are contained in 16-pin dual-in-line plastic packages. Figure 10-1A shows the top view of the MN6040 and MN6040A. Figure 10-1B shows the top view of the MN6040B. Table 10-1 shows the absolute maximum ratings of the MN6040 series, while Table 10-2 shows the recommended operating conditions of the MN6040 series. Table 10-3 lists the complete electrical specifications of the MN6040 series.

SPECIAL FEATURES

- ☐ Low-power CMOS construction
- ☐ Binary programming
- ☐ On-chip pulldown resistors on the program inputs
- ☐ Lock-detector output
- ☐ On-chip oscillator (MN6040A and MN6040B)
- ☐ Selectable reference frequency division (MN6040A)
- ☐ 5.12 MHz (one-half reference oscillator frequency) output (MN6040B)

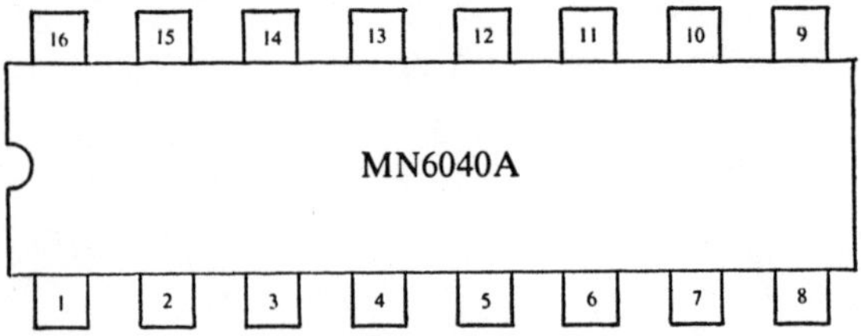

A

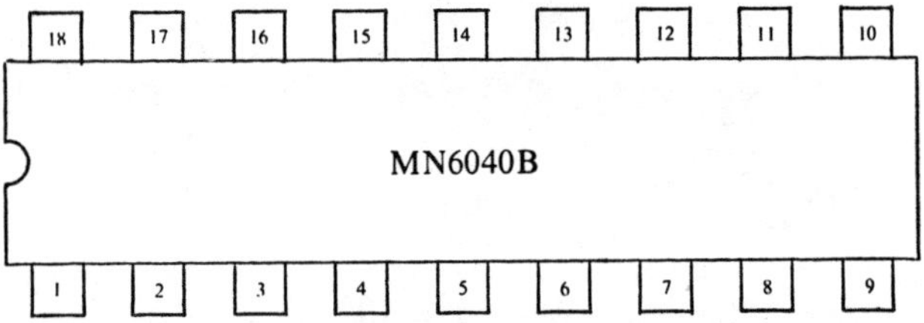

B

Fig. 10-1. Top view of the MN6040A PLL IC at A, and top view of the MN60-40B PLL IC at B.

Table 10-1. Absolute Maximum Ratings of the MN6040 Series (courtesy of Panasonic).

ABSOLUTE MAXIMUM RATINGS (Ta=25° C unless otherwise specified)

Item	Symbol	Rating	Unit
Supply Voltage	VDD	-0.3 +7.0	V
Input Voltage	VIN	-0.3 VDD+0.3	V
Output Voltage	VOUT	-0.3 VDD+0.3	V
Operating Ambient Temperature	Topr	-30 +70	°C
Storage Temperature	Tstg	-40 +100	°C

Table 10-2. Recommended Operating Conditions of the MN6040 Series (courtesy of Panasonic).

Item	Symbol	Conditions	Min	Typ	Max	Unit
Supply Voltage	VDD	VSS=OV, Ta=25° C	4.7	5.2	5.8	V

Table 10-3. Electrical Characteristics of the MN6040 Series (courtesy of Panasonic).

ELECTRICAL CHARACTERISTICS (V_{DD}=5.2V, V_{SS}=0V, f_{Q1}=10.24MHz, f_{P1}=2.55MHz, Ta=25°C)

Item	Symbol	Conditions	Min.	Typ.	Max.	Unit
Supply Current	I_{DD}	DS terminal in "H" Level			5	mA
Power Consumption	P_D	DS terminal in "H" Level			25	mW
PI Input Terminal						
Input Signal Frequency Upper Limit	f_{PI}	Input sine wave	2.55			MHz
Input Signal Swing	V_{PI}	Input sine wave	1.0			V_{P-P}
Feedback Resistance	R_{F1}		0.2			MΩ
QI Input Terminal						
Input Signal Frequency Upper Limit	f_{QI}	Input sine wave	10.24			MHz
Input Signal Swing	V_{QI}	Input sine wave	3.0			V_{P-P}
Feedback Resistance	R_{F2}		0.2			MΩ
DS Input Terminal (Except MN6040B)						
Input Voltage "H"	V_{IH1}		4.2		5.2	V
Input Voltage "L"	V_{IL1}		0		0.4	V
Input Current	I_{IN1}	V_{IN} = 0V	−0.15	−0.6	−1.2	mA
Input Leakage Current	I_{LK1}	V_{IN} = 5.2V			10	μA
P_0~P_8 Input Terminal (P_0~P_7 for MN6040A)						
Input Voltage "H"	V_{IH2}		3.7		5.2	V
Input Voltage "L"	V_{IL2}		0		0.4	V
Input Current	I_{IN2}	V_{IN} = 5.2V	0.15	0.6	1.2	mA
Input Leakage Current	I_{LK2}	V_{IN} = 0V			−10	μA
PD Output Terminal						
Output Current "H"	I_{OH1}	Vout = 3.2V	−0.8			mA
Output Current L	I_{OL1}	Vout = 2.0V	0.8			mA
Output Leakage Current "H"	I_{LO1H}	Vout = 5.2V			10	μA
Output Leakage Current "L"	I_{LO1L}	Vout = 0V			−20	μA
LD Output Terminal						
Output Current "H"	I_{OH2}	Vout = 3.2V	−0.8			mA
Output Current "L"	I_{OL2}	Vout = 2.0V	0.8			mA
Output Leakage Current "H"	I_{LO2H}	Vout = 5.2V			100	μA
Output Leakage Current "L"	I_{LO2L}	Vout = 0V			−20	μA
Q/2 Output Terminal (for MN6040B only)						
Output Current "H"	I_{OH}	Vout = 3.2V	−0.5			mA
Output Current "L"	I_{OL}	Vout = 2V	0.5			mA
Output Leakage Current "H"	I_{LOH}	Vout = 5.2V			10	μA
Output Leakage Current "L"	I_{LOL}	Vout = 0V			−20	μA

INTERNAL STRUCTURE

The functional block diagram of the MN6040 is shown in Fig. 10-2A. The various pin assignments of the MN6040 are as follows:

☐ Pin 1. This is connected to +5.2 volts supply source.

☐ Pin 2. This is the input to the programmable divider.

☐ Pin 3. This is the input to the reference divider. This divider divides by 1024, providing pin 4 is open or HIGH.

☐ Pin 4. When pin 4 is open or HIGH, the reference divider is enabled to divide by 1024.

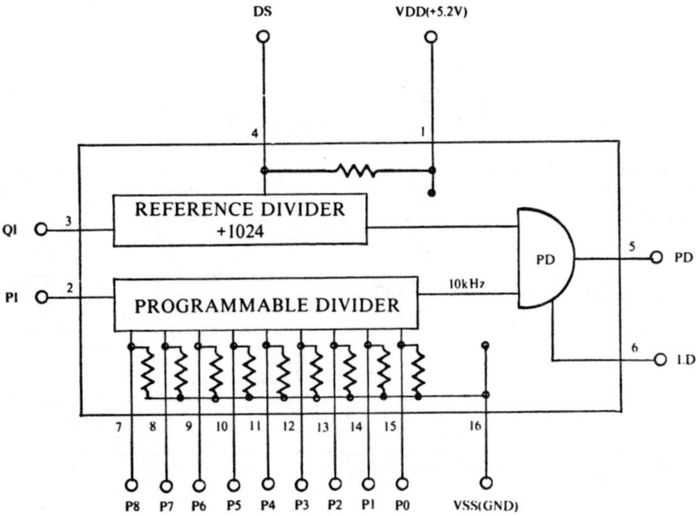

MN6040

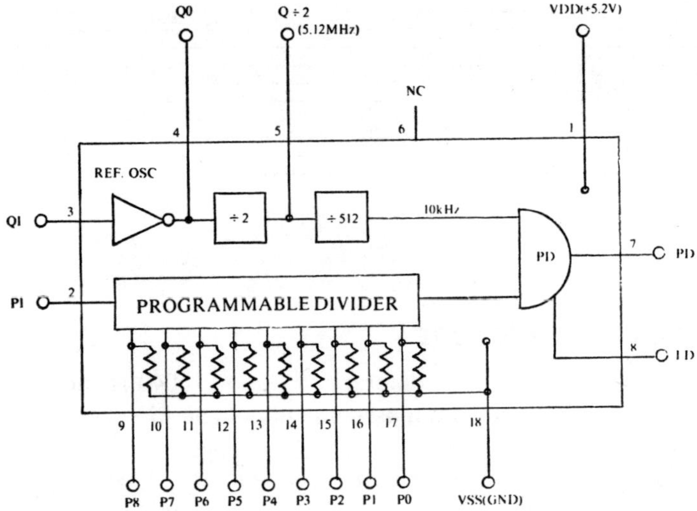

MN6040B

Fig. 10-2. A functional block diagram of the MN6040 at A, and a functional block diagram of the MN6040B at B (courtesy of Panasonic).

☐ Pin 5. This is the output of the phase detector. When the output of the programmable divider is greater than 10 kHz, PD is at the VSS level. When the output of the programmable divider is less than 10 kHz, PD is at the VDD level. When the output of the programmable divider is equal to 10 kHz, PD is open.

☐ Pin 6. This is the output of the lock detector (LD). Under normal, locked conditions, LD is at the HIGH level. When the loop is unlocked, LD provides randon output pulses.

☐ Pins 7 through 15. These are the programming inputs to the programmable divider. The pure binary code is used for programming, with P8 as the most significant bit while P0 is the least significant bit. Pulldown resistors are installed on each program input.

☐ Pin 16. This is the ground terminal for the IC.

The functional block diagram of the MN6040B is shown in Fig. 10-2B. The various pin functions are as follows:

☐ Pin 1. This connects to +5.2 volts to supply power to the IC.

☐ Pin 2. This is the input to the programmable divider.

☐ Pin 3. This is the input to the on-chip reference oscillator.

☐ Pin 4. This is the output of the on-chip reference oscillator.

☐ Pin 5. This is one-half the frequency of the reference oscillator. For example, with a 10.240-MHz reference oscillator, the signal at pin 5 will be 5.12 MHz.

☐ Pin 6. No internal connection is made to this pin.

☐ Pin 7. This is the output of the phase detector. It operates like the MN6040.

☐ Pin 8. This is the lock detector output. It too operates like the MN6040.

☐ Pins 9 through 17. These are the programming inputs to the programmable divider. P8, with a weight of 255, is the most significant bit (MSB). P0, with a weight of 1, is the least significant bit (LSB).

☐ Pin 18. This is the ground terminal of the IC.

The functional block diagram of the MN6040A is shown in Fig. 10-3. The various pin assignments are as follows:

☐ Pin 1. This is the supply voltage to the IC.

☐ Pin 2. This is the input to the programmable divider. The specifications call for a maximum frequency of 2.55 MHz at this input.

☐ Pin 3. This is the input side of the on-chip reference oscillator.

☐ Pin 4. This is the output side of the on-chip reference oscil-

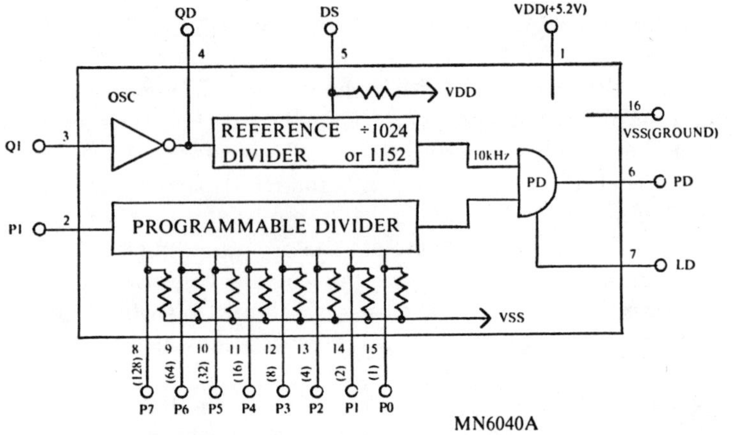

Fig. 10-3. A functional block diagram of the MN6040A (courtesy of Pana sonic).

lator. The reference frequency crystal would connect between pins 3 and 4.

☐ Pin 5. This is the reference divider select pin. If DS (pin 5) is in the HIGH level, the reference divider is set to divide by 1024. If this pin is at the LOW level (ground), the reference divider is set to divide by 1152. The frequency of the reference oscillator would depend upon the divide ratio of the reference divider. If the reference divider ratio is selected as 1024, the reference frequency should then be 10.24 MHz. If the ratio is selected as 1152, the reference oscillator frequency should then be 11.52 MHz. There is a pullup resistor connected to the DS terminal, so if this pin is left open, the divide ratio is 1024.

☐ Pin 6. This is the error signal output from the phase detector. If the output from the programmable divider is greater than 10 kHz, pin 6 goes to a LOW level (VSS). If the output from the programmable divider is less than 10 kHz, pin 6 goes to a HIGH level. If the output from the programmable divider is equal to 10 kHz, pin 6 is open.

☐ Pin 7. This is the output of the lock detector. During normal, locked conditions, pin 7 is at a HIGH level. If the loop becomes unlocked, the LD output provides random output pulses.

☐ Pins 8 through 15. These are the programming inputs to the programmable divider. Programming is done using the pure binary code. Pin 8 (P7), with a weight of 128, is the most significant bit (MSB) while pin 15 (P0), with a weight of 1, is the least significant bit (LSB). Each of the program lines (P0 through P7) has an on-chip pulldown resistor. If any of these pins are left open, the pulldown resistor places the pin at binary 0.

☐ Pin 16. This is the ground pin of the IC.

ROBYN SX-402D PLL SYNTHESIZER

The complete schematic diagram of the PLL synthesizer of the Robyn SX-402D CB set is shown in Fig. 10-4. This synthesizer requires the use of three crystals. The function of the various stages of this PLL circuit are as follows:

PLL Synthesizer IC

First, check the programming pins (8 through 15). Pin 8, with a weight of 128, is connected to VDD (+ 5.2 volts). Thus, 128 will always be used in determining the N divisor. Since pin 9 is connected to ground, its weight of 64 will never be used in determining the N divisor. All the other program pins (10 through 15) are connected to the channel selector switch. The channel selector switch applies binary 1 (+ 5.2 volts) to the appropriate pins. The pins which are to be at the binary 0 level are left open. The pulldown resistors pull these pins down to binary 0.

The reference divider ratio selector (pin 5), DS, is left open. The pullup resistor on the pin places it at the binary 1 level. This selects a divide ratio of 1024. The reference oscillator will therefore operate at 10.240 MHz. The on-chip reference oscillator is not used as the oscillator. Instead, a separate oscillator is used to generate the 10.240-MHz reference frequency. However, the on-chip amplifier is used as a buffer for the reference oscillator signal.

Reference Oscillator and Delta-Tune

The reference oscillator consists of transistor TR1 and its associated circuitry. The frequency of the oscillator is primarily determined by crystal X1, which resonates at 10.240 MHz. Tuning coil L1 can be used to make the frequency adjustments. Also, associated with the reference oscillator circuit is the delta-tune feature. Notice that crystal X1 connects to ground through varactor diode D2. In the receive mode, diode D3 is forward biased. The positive voltage at the arm of the delta-tune control, VR5 is applied to the cathode of the varactor through D3 and R51. This reverse biases the varactor. By varying the reverse bias to the varactor, the frequency of the reference oscillator can be changed

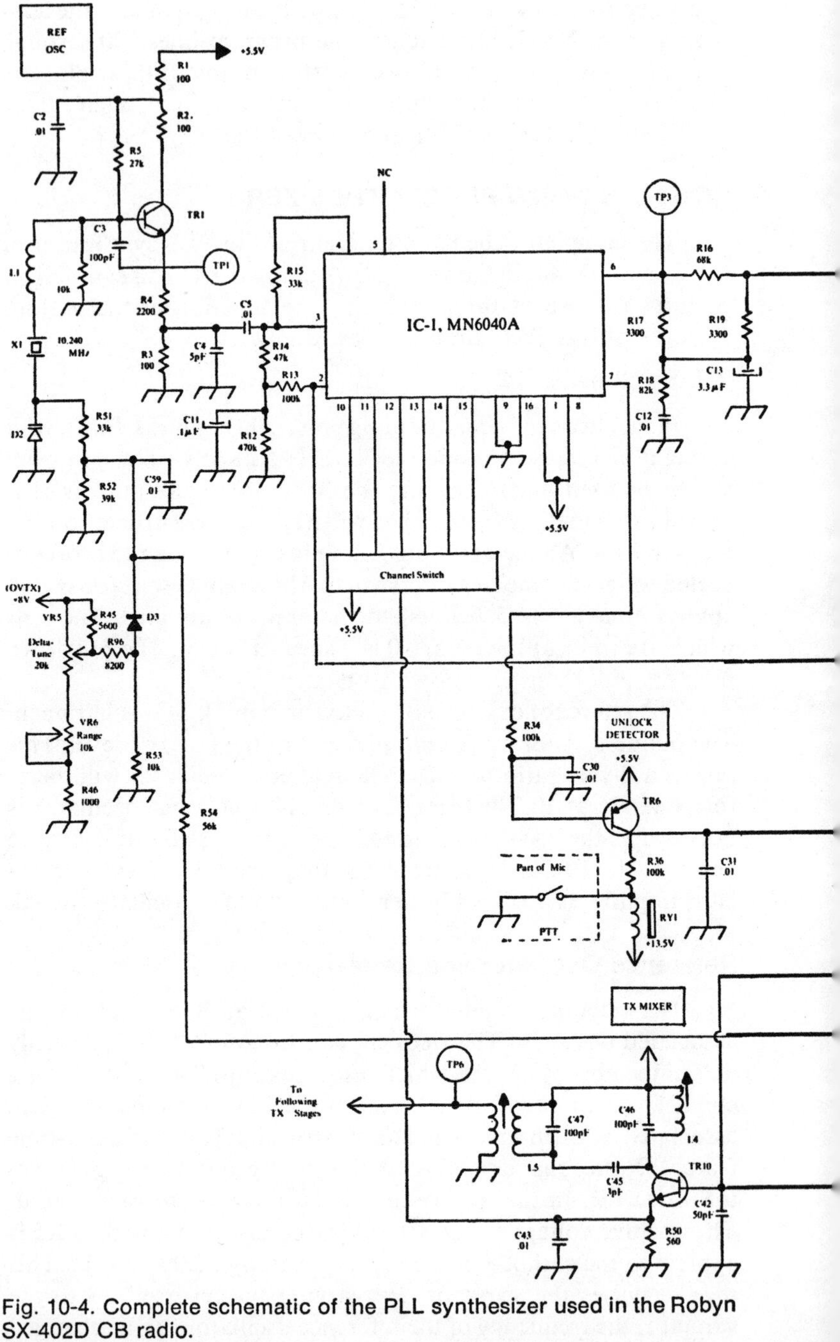

Fig. 10-4. Complete schematic of the PLL synthesizer used in the Robyn SX-402D CB radio.

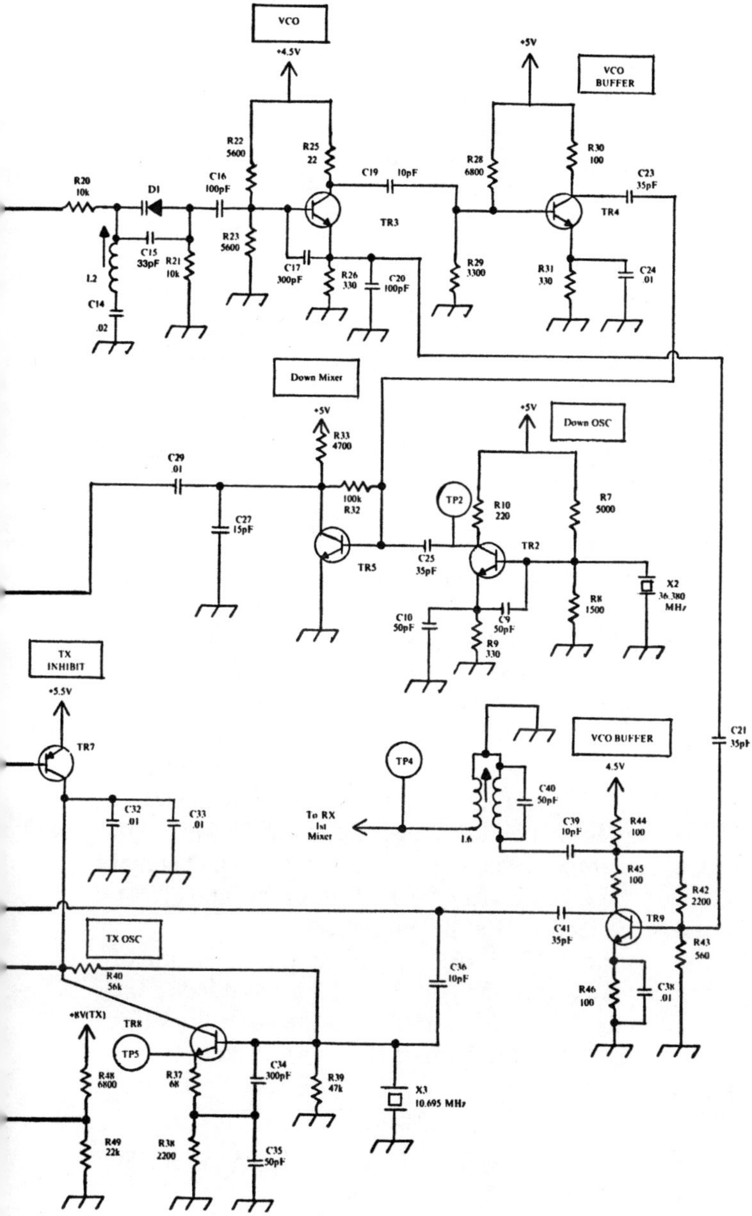
VCO
+4.5V
+5V
VCO
BUFFER
R22
5600
R25
22
R30
100
R28
6800
C23
35pF
R20
10k
D1
C16
100pF
C19
10pF
TR3
TR4
R23
5600
C15
33pF
R21
10k
L2
C14
.02
C17
300pF
R26
330
C20
100pF
R29
3300
R31
330
C24
.01
Down Mixer
+5V
R33
4700
+5V
Down OSC
C29
.01
100k
R32
TP2
R10
220
R7
5000
C27
15pF
TR5
C25
35pF
TR2
X2
36.380
MHz
R8
1500
C10
50pF
C9
50pF
R9
330
TX
INHIBIT
+5.5V
TR7
VCO BUFFER
C21
35pF
TP4
4.5V
C40
50pF
C32
.01
C33
.01
To RX
1st
Mixer
L6
C39
10pF
R44
100
R45
100
R42
2200
TX OSC
C41
35pF
TR9
R43
560
R40
56k
C36
10pF
R46
100
C38
.01
+8V(TX)
TR8
TP5
R48
6800
R37
68
C34
300pF
R39
47k
X3
10.695 MHz
R49
22k
R38
2200
C35
50pF

within a given range. This is what the delta-tune control does.

In the transmit mode, the supply voltage at the top of the delta-tune control is removed and a positive voltage appears at the collector of TR8 (transmit oscillator). Thus diode D3 is now reverse biased, effectively disconnecting the delta-tune control from the circuit. The reverse bias on the cathode of varactor D2 is set at a fixed level by the voltage divider consisting of R52 and R54. This bias sets the reference oscillator at its center frequency of 10.240 MHz.

VCO and Buffers

The VCO oscillator is built around transistor TR3. Feedback is provided by capacitor C17 between the base and emitter of TR3. The VCO operates in the 37-MHz range. The resonant frequency of the VCO is determined by L2, C15, D1, C16. These components comprise a series resonant circuit. L2 can be used to set the VCO to the proper frequency range. Fine adjustment in VCO frequency is then made by the control voltage, which is connected to the cathode of varactor D1. The voltage at the cathode of D1 will be a positive value. Because the anode of D1 is grounded through R21, the positive voltage on the cathode will reverse bias D1. If the positive voltage on the cathode will increase, the capacitance exhibited by varactor D1 will decrease, thus increasing the VCO frequency. Conversely, if the positive voltage on the cathode decreases, the capacitance exhibited by D1 will increase, thus decreasing the VCO frequency.

The VCO signal is fed from the emitter of TR3 through C21 to the base of VCO buffer, TR9. This stage amplifies the VCO signal. A VCO signal is then fed from the collector of TR9 through C41 to the transmit mixer. Also, a portion of the VCO signal appears at the secondary of transformer L6 at test point 4 (TP4). This signal feeds the receive first mixer stage.

Going back to the VCO (TR3), notice also that a portion of the VCO signal is taken from the collector of TR3 and fed to the base of TR4 through C19. VCO buffer, TR4 then amplifies this VCO signal and sends it to the base of the down mixer, TR5.

Down Oscillator

The down oscillator provides a fixed frequency of 36.380 MHz, which is mixed with the VCO frequency to produce an offset frequency fed to the programmable divider. The down oscillator frequency is controlled by crystal X2. Notice that the collector of TR2 is TP2. The down oscillator signal is fed from the collector of TR2 to the base of the down mixer (TR5) through coupling capacitor C25.

Down Mixer

The down mixer (TR5) is a simple common-emitter amplifier. Two signals are fed to the base, one from the down oscillator and the other from VCO buffer TR4. The down mixer mixes the two frequencies. The difference between the two frequencies is then fed from the collector to the input of the programmable divider.

Transmit Oscillator

The transmit oscillator is active only in the transmit mode. This is a crystal-controlled oscillator that generates a 10.695-MHz signal used to mix with the VCO signal to provide a difference frequency. This is the transmit frequency. Crystal X3 resonates the oscillator at 10.695 MHz. The emitter transmit oscillator TR8 is TP5.

Transmit Mixer

The transmit mixer, TR10, mixes the VCO frequency with the 10.695-MHz transmit oscillator frequency to provide a difference frequency. This is the transmit frequency. The tuned circuit in the collector resonates at the transmit frequency. The other frequencies are attenuated.

Unlock Detector and Transmit Inhibit

In the transmit mode, if the lock detector output at pin 7 of IC1 should go LOW (as it will if the loop is unlocked), transistor TR6 will be forward biased. This places a high positive voltage on the base of TR7, causing TR7 to cut off. Because the supply voltage to the transmit oscillator (TR8) is fed through TR7, when TR7 cuts off, this kills the transmit oscillator and the transmitter output.

In the receive mode, TR7 is cut off because a high positive voltage is applied through relay coil RY1 and R36 to the base of TR7, reverse biasing TR7. When the transmit push-to-talk button is pressed, the junction of the relay coil and R36 is grounded. This forward biases TR7, thus energizing the transmit oscillator.

Notice on the schematic that a wire connects from the channel selector switch to the emitter of TR10, the transmit mixer. If the channel selector is placed in a position between channels, this places a high positive voltage on the emitter of TR10. This reverse biases TR10, thus disabling the transmitter output. The reason for doing this is because improper (in-between channel) settings of the channel selector could cause improper programming to be applied to the programmable divider. This could cause the transmitter to work on an unauthorized frequency.

LED Channel Display Circuit

The LED channel display circuit is shown in Fig. 10-5A. The rear view of this display is shown at Figure 10-5B. This display consists of two seven-segment LEDs on a single block. One display serves as the tens digit and the other as the units digit. The display is a common-anode type, since the common pin (5) is supplied with positive voltage. The cathodes are grounded individually through contacts on the channel selector switch. Notice, however, that on the tens digit display, segments a and d are connected together. This is because they are always on or off at the same time so there is no need to switch them individually. If you check the segments which are energized for numbers 1 through 4, you will see that segments a and b are on or off simultaneously. Since the positive voltage supplied to the common-anode point (pin 5) is only 1.9 volts, there is no need for individual current-limiting resistors in the cathode lines.

Circuit Analysis

When channel 40 is selected, the N value programmed into the programmable divider is 172. Remember that pin 8, with a weight of 128, is always at binary 1 and pin 9, with a weight of 64, is always at binary 0. The other program pins (10 through 15) connect to the channel switch, which determines their binary levels for each channel. The total N number will be equal to 128 plus the decimal value programmed into pins 10 through 15. Table 10-4 shows the N number, programming code, and VCO frequency for channels 1 through 40. The weight of program pins 10 through 15 are shown above the pin number. At channel 40, the table shows that the binary 1 appears on pins 10, 12, and 13. The sum of the weights of these pins is 32 + 8 + 4, or 44. Adding 44 to 128 gives the N value of 172.

The correct VCO frequency required for channel 40 is 38.100 MHz. When the radio is first switched on, however, the VCO fre-

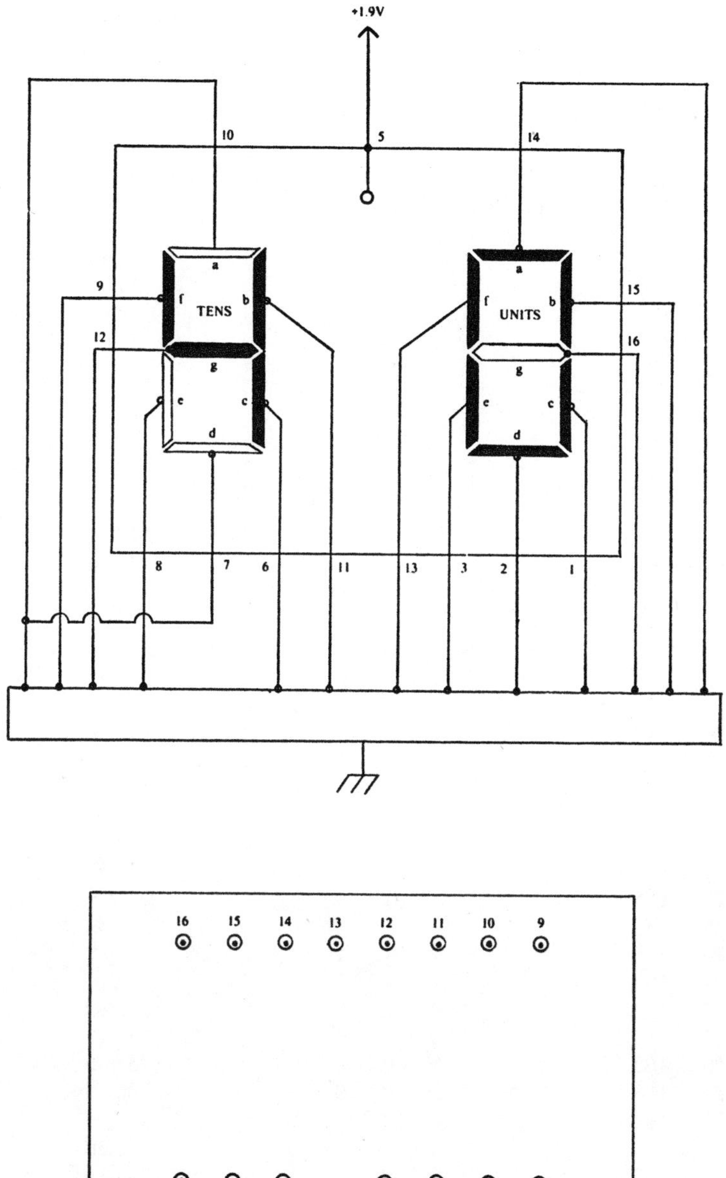

Fig. 10-5. The LED channel display circuit used in the Robyn SX-402D at A, and a rear view of the LED channel display used in the Robyn SX-402D that shows the pin numbering scheme.

Table 10-4. Robyn SX-402D PLL Synthesizer Specifications.

CH.	N	PRG CD-PN 10-15						VCO FREQ.
		Pin weight---32	16	8	4	2	1	AT TP4 (MHz)
		Pin no.-------10	11	12	13	14	15	
1	128	0	0	0	0	0	0	37.660
2	129	0	0	0	0	0	1	37.670
3	130	0	0	0	0	1	0	37.680
4	132	0	0	0	1	0	0	37.700
5	133	0	0	0	1	0	1	37.710
6	134	0	0	0	1	1	0	37.720
7	135	0	0	0	1	1	1	37.730
8	137	0	0	1	0	0	1	37.750
9	138	0	0	1	0	1	0	37.760
10	139	0	0	1	0	1	1	37.770
11	140	0	0	1	1	0	0	37.780
12	142	0	0	1	1	1	0	37.800
13	143	0	0	1	1	1	1	37.810
14	144	0	1	0	0	0	0	37.820
15	145	0	1	0	0	0	1	37.830
16	147	0	1	0	0	1	1	37.850
17	148	0	1	0	1	0	0	37.860
18	149	0	1	0	1	0	1	37.870
19	150	0	1	0	1	1	0	37.880
20	152	0	1	1	0	0	0	37.900
21	153	0	1	1	0	0	1	37.910
22	154	0	1	1	0	1	0	37.920
23	157	0	1	1	1	0	1	37.950
24	155	0	1	1	0	1	1	37.930
25	156	0	1	1	1	0	0	37.940
26	158	0	1	1	1	1	0	37.960
27	159	0	1	1	1	1	1	37.970
28	160	1	0	0	0	0	0	37.980
29	161	1	0	0	0	0	1	37.990
30	162	1	0	0	0	1	0	38.000
31	163	1	0	0	0	1	1	38.010
32	164	1	0	0	1	0	0	38.020
33	165	1	0	0	1	0	1	38.030
34	166	1	0	0	1	1	0	38.040
35	167	1	0	0	1	1	1	38.050
36	168	1	0	1	0	0	0	38.060
37	169	1	0	1	0	0	1	38.070
38	170	1	0	1	0	1	0	38.080
39	171	1	0	1	0	1	1	38.090
40	172	1	0	1	1	0	0	38.100

quency will not be exactly 38.100 MHz. If the VCO frequency is higher than this, the output of the programmable divider will be greater than 10 kHz. The phase detector output will then go to a lower positive level. This lower positive level is applied to the VCO control line, thus reducing the reverse bias on the varactor D1. This causes the capacitance of the varactor to increase, thus decreasing the VCO frequency. When the VCO frequency reaches 38.100 MHz, the input to the programmable divider will be 38.100 MHz - 36.380 MHz, or 1.72 MHz. The output of the programmable divider will then be 1.72 MHz ÷ 172, or 10 kHz. Since this is equal to the reference frequency, the phase detector error signal will disappear. The control voltage applied to the VCO control line will contain little or no error component so the VCO frequency stabilizes at 38.100 MHz.

In the receive mode, this 38.100-MHz VCO signal is fed from VCO buffer TR9 through L6 to the receive first mixer, where it is heterodyned with the incoming receive signals. Since the output of the first mixer is tuned to 10.695 MHz, the receive frequency will be 38.100 MHz - 10.695 MHz, or 27.405 MHz, which is channel 40. This 10.695-MHz first i-f is then fed to the receive second mixer, where it is heterodyned with a 10.240-MHz signal from the reference oscillator. This yields a difference frequency of 10.695 MHz - 10.240 MHz, or 455 kHz, the second intermediate frequency.

In the transmit mode, the 38.100-MHz VCO signal is fed from the collector of VCO buffer TR9 to the base of the transmit mixer, TR10. Here, it is mixed with a 10.695-MHz signal from the transmit oscillator, TR8. A difference frequency of 38.100 MHz - 10.695 MHz, or 27.405 MHz, appears at the output of the receive mixer. This is then fed to the following transmit stages for further amplification.

Alignment

☐ Set the delta-tune control to midposition and connect a frequency counter to TP1, the emitter of TR1. Check for 10.240 MHz. If the frequency is incorrect, adjust L1 for exactly 10.240 MHz.

☐ Connect a frequency counter to TP2, the collector of TR2, and check for 36.380 MHz.

☐ Set the delta-tune control to midposition and set the channel selector to channel 1. Connect a DC voltmeter to TP3 and adjust L2 for 2.3 volts at TP3.

☐ Set the channel selector to channel 20, connect an rf voltmeter to TP4, and adjust L6 for maximum indication on the rf voltmeter.

☐ Set the delta-tune control to midposition. Check for proper VCO frequency at TP4 for each channel. Refer to Table 10-4 for the proper VCO frequency for each channel.

☐ Connect a frequency counter to TP5, the emitter of TR9. Key the transmitter and check for 10.695 MHz.

☐ Set the channel selector to channel 20. Connect an rf voltmeter to TP6. Key the transmitter and adjust L4 and L5 for maximum indication on the rf voltmeter.

☐ Connect a frequency counter to TP6. Key the transmitter and check for proper transmit frequency on all channels.

Troubleshooting

What do you do when the rig neither transmits nor receives? If the PLL section is responsible for this trouble, the VCO signal at TP4 will probably be missing altogether or will be far off frequency. If the VCO signal is missing completely, troubleshoot the VCO and VCO buffer TR9. If the VCO signal is present at TP4 but is far off frequency, check the lock detector output voltage at pin 7 of IC1. If the voltage there is low, this indicates that the loop is unlocked. Check for the proper 10.240-MHz reference signal at pin 3 of IC1. If this signal is missing or is far off frequency, troubleshoot the reference oscillator. Check for the offset frequency signal. If this is missing, use your rf probe to signal trace to the collector of TR5. Check for rf voltage at TP2 and use a frequency counter to verify that the frequency at TP2 is 36.380 MHz. Check for the VCO signal at TR4 collector. If the offset signal is present at pin 2 of IC1 but the frequency is far off and unstable, then the trouble is probably with the VCO or VCO control line. The voltage at pin 6 of IC1 will probably be abnormally high or abnormally low. It is possible that the VCO has been misadjusted or is out of alignment for some other reason. Monitor the DC voltage at pin 7 of IC1 (the lock detector output) while carefully and slowly tuning coil L2. If the voltage at pin 7 of IC1 rises to near VDD level, move the DC meter probe to TP3 and adjust L2 for 2.3 volts on channel 1. If tuning L2 has no effect on the lock detector voltage at pin 7, return the core of L2 to its original position.

It is also possible that the MN6040A (IC1) may have some internal defect which could cause the problem. Sometimes, the only way to prove or disprove that the IC is bad is to substitute a new one, but only do this as a last resort.

High Channels Inoperative with Low Channels OK. This is usually caused by VCO misalignment. The voltage at TP3 will probably be abnormally high. See the alignment section for proper adjustment of the VCO. If this isn't the problem, check the frequency determining components of the VCO.

Low Channels Inoperative with High Channels OK. This is usually caused by VCO misalignment. the voltage at TP3 will probably be abnormally low. See the alignment section for proper adjustment of the VCO. If this isn't the trouble, check the frequency-determining circuit of the VCO.

Wrong Transmit Frequencies on All Channels. Check the 10.695-MHz oscillator by connecting a frequency counter to TP5. The 36.380-MHz down oscillator could affect the transmit frequency also, but the receive frequency would also be affected. You can check for the proper 36.380-MHz signal frequency at TP2.

Wrong Receive and Transmit Frequencies on All Channels. Check the frequency of the 36.380-MHz down oscillator at TP2. Also, check the 10.240-MHz reference frequency at TP1.

No Transmit Frequency Signal with Receive OK. Check the transmit oscillator and the transmit mixer. Also, check the transmit inhibit circuit.

No Receive with Transmit OK. Check at TP4 for proper signal. The proper signal must be at the collector of TR9 because the transmit signal is OK. Check C39 and L6.

Some Channels Inoperative. Check for proper programming to IC1. Check the channel switch.. Check for possible open or shorted program lines. This could be caused by a defective IC1 in some cases.

Some Channels Intermittent. This is most often caused by loose or dirty channel selector contacts. Clean or replace as necessary.

Chapter 11
Toshiba PLL ICs

The TC5080P, TC5081P, and TC5082P are a group of PLL IC devices designed to be used together to form the heart of a PLL synthesizer. These devices are manufactured by Toshiba. Each of these devices are discussed here.

TC5080P

The TC5080P is a CMOS programmable divider. This device is housed in a 16-pin dual-in-line plastic package. The outline of the TC5080P is shown in Fig. 11-1. A simplified functional block diagram of the TC5080P is shown in Fig. 11-2. Table 11-1 lists the maximum ratings of the TC5080P. The TC5080P can be programmed in binary form for any N number from 8 to 255. The TC5080P can also be programmed in BCD form for any N number from 8 to 99. The maximum input frequency which the TC5080P can handle is listed as 2.4 MHz. The complete electrical characteristics of the TC5080P are given in Table 11-2. The values listed in Table 11-2 are for an ambient temperature range of -30 to +75 degrees celsius.

The test circuit for determining the frequency range of the TC5080P is shown in Fig. 11-3A. The voltage divider network consisting of resistors R1 and R2 supplies bias to the CMOS device. On some PLL ICs, bias is supplied internally, but most of these devices require external positive bias voltage on the clock input. This is why many PLL circuits use direct coupling from

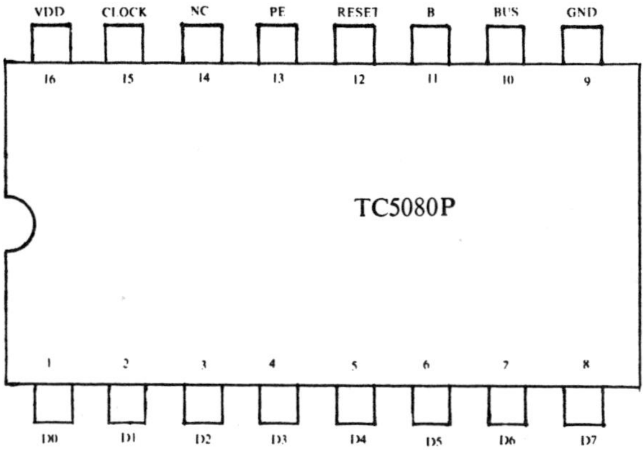

Fig. 11-1. Outline of the TC5080P showing pin numbers and functions (courtesy of Toshiba).

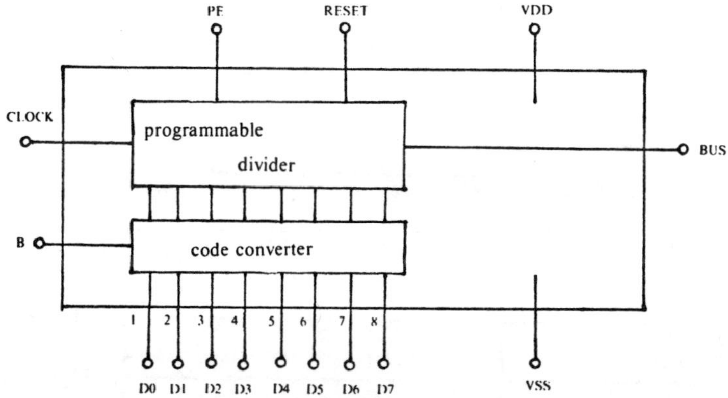

Fig. 11-2. Functional block diagram of the TC5080P.

the PLL down mixer to the input of the programmable divider. When capacitive coupling is used, you will usually find a resistor biasing network.

Figure 11-3B shows the minimum signal input voltage to the programmable divider. Notice that the rated minimum input p-p voltage is equal to ½ VDD.

Figure 11-3C shows the signal level required for clock frequencies from 1 kHz to 2.4 MHz. This can only be taken to mean that for operation at this clock frequency range, the input signal must swing from 0 volts to VDD level. The description of the various pin assignments are as follows:

Table 11-1. Absolute Maximum Ratings of the TC5080P (courtesy of Toshiba).

Ta = 25° C

CHARACTERISTIC	SYMBOL	RATING	UNIT
Supply voltage	VDD	10	V
Input voltage	Vin	-0.3 to VDD + 0.3	V
Operating tempera- ture	Top	-30 to +75° C	C
Storage tempera ture	Tstg	-55 to +125	°C

Table 11-12. Electrical Characteristics of the TC5080P (courtesy of Toshiba).

Ta = -30 to + 75° C

Characteristics	Symbol	Test Condition	Min.	Typ.	Max	Unit
Operating Supply voltage	VDD		4.5		8.0	V
Output voltage High level	VOH	VDD=7.5V, VIL=1.5V VIH=6.oV, IOL=50 uA	7.3			V
Output voltage low level	VOL	VDD=7.5V, VIL=1.5V VIH=6.0V, IOL=50uA			0.2	V
Quiescent current	IDD	VDD=7.5V, VIL=1.5V VIL = OV			500	uA
Maximum clock fre- quency	fMAX	VDD=7.5V (Ta=25°C) Vin=½ VDD	2.4			MHz
Minimum Input voltage	Vin(min)		½ VDD			Vp-p
Operating frequency range	fIN	VDD=7.5V (Ta=25°C) Vin=0 - VDD	1 kHz to 2.4 MHz			

□ Pins 1-8. These are the programming inputs of the programmable divider. Pin 1 (D0) is the least significant bit (LSB) while pin 8 (D7) is the most significant bit (MSB). The pro-

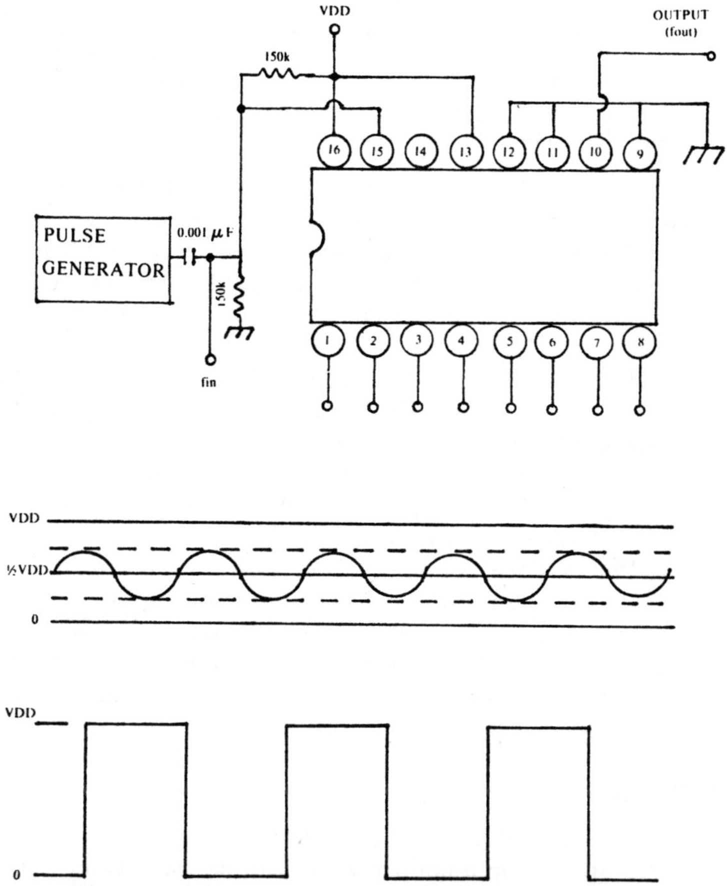

Fig. 11-3. Test circuit used for determining the frequency range of the TC5080P at A, the minimum signal input voltage for the TC5080P at B, and the signal level required for clock frequencies from 1 kHz to 2.4 MHz at C (courtesy of Toshiba).

gramming can be done in either the pure binary form or the BCD form. The type of programming used is determined by the voltage level at the code select terminal, pin 11. Table 11-3 shows the weight of each program pin for both the pure binary code and the BCD code. In the pure binary coding, the highest N value which can be programmed is 255. With BCD coding, the highest N value which can be programmed is 99. There are no pullup or pulldown resistors on these program inputs so external pullup or pulldown resistors are usually used.

☐ Pin 9. This is the ground terminal or VSS.

Table 11-3. TC5080P Programming Pin Weights.

Pin	1	2	3	4	5	6	7	8
Binary	1	2	4	8	16	32	64	128
BCD	1	2	4	8	10	20	40	80

□ Pin 10. This is the output of the programmable divider. The frequency of this output is equal to the clock frequency ÷ N.

□ Pin 11. This is the programming code select pin. If B is low (grounded), the programming is accepted in the pure binary form. If B is high (near VDD), then the programming is accepted in the BCD form.

□ Pin 12. When the RESET terminal is LOW (ground) the counter counts. If the RESET terminal is HIGH (near VDD), all stages of the counter are cleared.

□ Pin 13. When this terminal is HIGH (near VDD), the counter is enabled to operate. If the PE terminal goes LOW (ground), the counter is disabled.

□ Pin 14. There is no connection made to this terminal pin.

□ Pin 15. This is the input to the programmable divider. This terminal must be supplied with an external positive bias.

□ Pin 16. This is the supply voltage to the IC. The voltage here can range from 4.5 volts to 8.0 volts. The absolute maximum voltage is listed as 10 volts.

TC5081P

The TC5081P is a CMOS phase comparator. The device also has an on-chip amplifier which can be used with external components to form an active filter. The TC5081P is housed in a 9-pin single-in-line plastic package. The outline of the TC5081P is shown in Fig. 11-4. The functional logic diagram of the TC5081P is shown in Fig. 11-5. The timing diagram of the phase comparator is shown in Fig. 11-6. The S input is normally the reference frequency input while the R input is the frequency to be compared, such as the output of the programmable divider. If the frequency at the R input is higher than the frequency at the S input, the resulting pulses at PD out will be positive-going. If the frequency at the R input is lower than the reference frequency, the resulting pulses at PD out will be negative-going. When the S and R frequencies are the same, PD out will go to a high impedance or third state. When the amplifier is used as part of an active filter, the output at the A out is an inversion of the input. So if the aver-

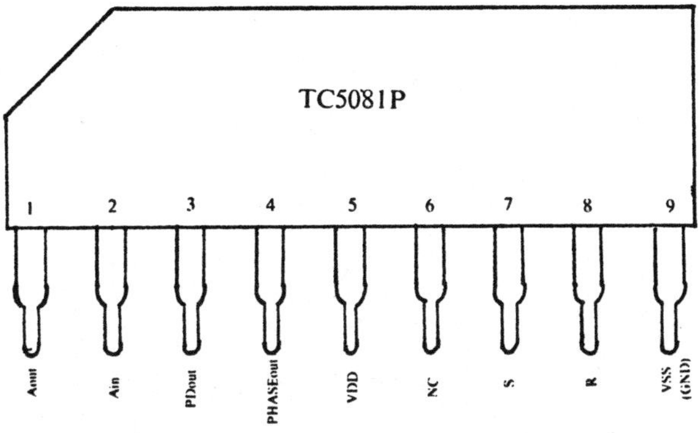

Fig. 11-4. Outline of the TC5081P showing pin numbers and functions (courtesy of Toshiba).

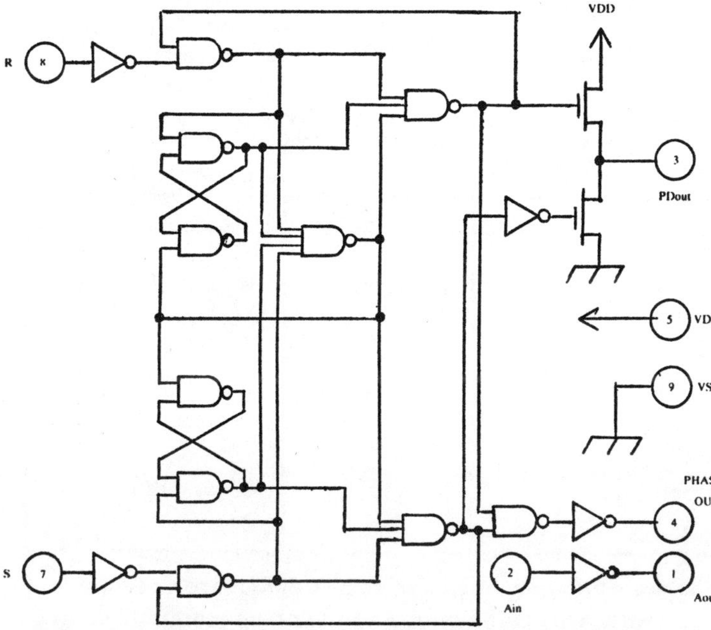

Fig 11-5. Logic diagram of the TC5081P (courtesy of Toshiba).

age positive DC level at PD out increases, the average positive DC level at A out will decrease. And if the average positive DC voltage level at PD out decreases, the average positive level at A out will increase. The voltage from A out is the control voltage which is applied to the VCO control line to steer the VCO frequency.

The absolute maximum ratings of the TC5081P are listed in Table 11-4. The electrical characteristics are listed in Table 11-5.

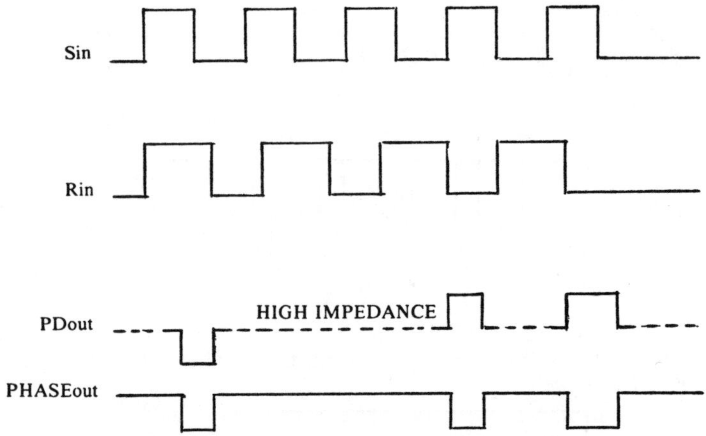

Fig. 11-6. Timing chart for the TC5081P (courtesy of Toshiba).

Table 11-4. Absolute Maximum Ratings of the TC5081P (courtesy of Toshiba).

MAXIMUM RATINGS (Ta = 25 Degrees C)

Characteristics	Symbol	Rating	Unit
Supply Voltage	VDD	10	V
Input voltage	Vin	-0.3 to VDD + 0.3	V
Operating temperature	Top	-30 to +75 Degrees	C
Storage temperature	Tstg	-55 to +125 Degrees	C

The various pin assignments of the TC5081P are as follows:

- ☐ Pin 1. This is the output of the on-chip amplifier.
- ☐ Pin 2. This is the input to the on-chip amplifier.
- ☐ Pin 3. This is the error signal output from the charge pump.
- ☐ Pin 4. This is the output of the lock detector. Normally (under locked loop conditions) this terminal is at or near the VDD level. If the loop becomes unlocked, the output from this pin is a series of negative going pulses.
- ☐ Pin 5. This is the supply voltage input to the IC.
- ☐ Pin 6. There is no internal connection to this pin.
- ☐ Pin 7. This is the set input. Normally the reference signal is

Table 11-5. Electrical Characteristics of the TC5081P Phase Comparator IC (courtesy of Toshiba).

ELECTRICAL CHARACTERISTICS (Ta = -35 to +75° C)

Characteristic	Symbol	Test Condition	MIN	TYP	MAX	UNIT
Operating supply voltage	VDD		4.5		8.0	V
Output voltage high level	VOH	VDD=7.5V, VIL=1.6V VIH=6.6V, IOH= -50 uA	7.3			V
Output voltage low level	VOL	VDD=7.5V, VIL=1.6V VIH=6.6V, IOL=50 uA			0.2	V
Quiescent current	IDD	VDD=7.5V, VIH=7.5V, VIL=OV			200	uA
3 State leak current	ITLH	VDD=7.5V			500	nA
	ITLL	VDD=7.5V			-500	nA

applied to this terminal.

☐ Pin 8. This is the reset input. Normally the signal which is to be compared is applied to this input.

☐ Pin 9. This is the ground terminal for the IC.

TC5082P

The TC5082P is a CMOS oscillator and 12-stage frequency divider. This device is housed in a 9-pin single-in-line plastic package. The outline drawing of the TC5082P is shown in Fig. 11-7. The maximum ratings of the TC5082P are listed in the chart in Table 11-6. Table 11-7 lists the electrical characteristics of the TC5082P. Figure 11-8 shows a drawing of the test circuit used to determine the electrical characteristics. Table 11-8 lists the pin assignments of the TC5082P. The functional block diagram of the TC5082P is shown in Fig. 11-9. Normally the reference frequency oscillator operates at 10.240 MHz, as governed by the external crystal. Table 11-9 shows the available output frequencies with the 10.240-MHz reference signal.

THE TRAM D12 PLL SYNTHESIZER

The Tram D12 PLL synthesizer is included here because it has some unusual features which make it quite interesting. A block diagram of the Tram D12 PLL synthesizer is shown in Fig. 11-10. First of all, notice that this synthesizer uses not one but two TC5081P phase comparators. Also, a voltage-controlled multivibrator is used to generate a frequency that is used as the input to

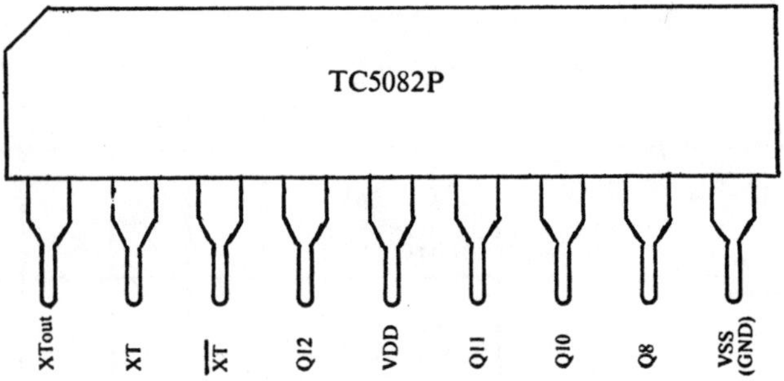

Fig. 11-7. Outline of the TC5082P showing pin numbers and functions (courtesy of Toshiba).

Table 11-6. Maximum Ratings of the TC5082P IC (courtesy of Toshiba).

MAXIMUM RATINGS (Ta = 25° C)

Characteristics	Symbol	Rating	Unit
Supply Voltage	VDD	10	V
Input Voltage	Vin	-0.3 to VDD + .3	V
Operating temperature	Top	-30 to +75°	C C
Storage temperature	Tstg	-55 to +125	°C

Table 11-7. Electrical Characteristics of the TC5082P (courtesy of Toshiba).

ELECTRICAL CHARACTERISTICS (Ta = -30 to +75° C)

Characteristic	Symbol	Test Condition	MIN	TYP	MAX	UNIT
Operating supply voltage	VDD	X'tal=10.24 MHz	4.5		8.0	V
Operating current	IDD	X'tal=10.24 MHz VDD=7.5 V			7.0	mA
Output voltage High level	VOH	VDD=7.5 V, IOH = -50 uA	7.3			V
Output voltage Low level	VOL	VDD=7.5 V, IOL = 50 uA			0.2	V
Maximum clock frequency	fMAX	VDD= 7.5 V	10.24			MHz
①Pin output voltage	Vout 1	VDD= 7.5 V, CL =15 pF X'tal = 10.24 MHz	3.5			V p-p

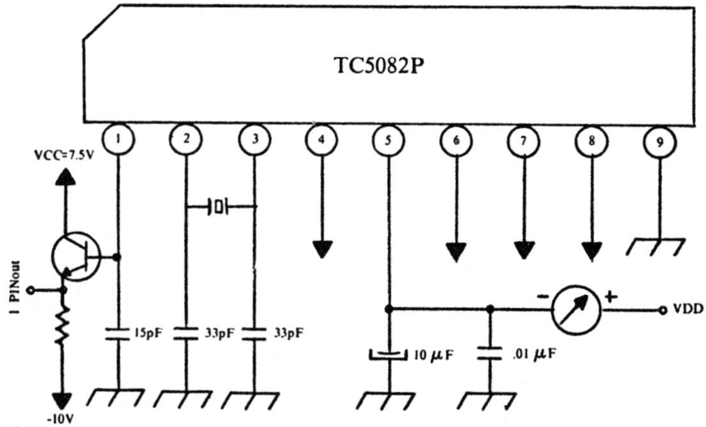

Fig. 11-8. Test circuit used to determine the electrical characteristics of the TC5082P (courtesy of Toshiba).

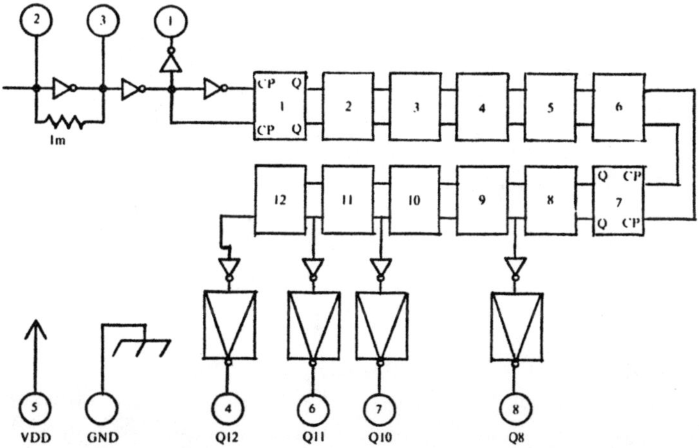

Fig. 11-9. Functional block diagram of the TC5082P (courtesy of Toshiba).

the programmable divider and as an input to one of the phase detectors. Another unusual feature of this synthesizer is that there is a very large shift in the VCO frequency between the receive transmit modes. Also, in the transmit mode the VCO operates at the transmit frequency.

The TC5082P (IC8) has an on-chip oscillator that operates at a frequency of 5.12 MHz, as governed by crystal X1. The output from pin 7 of IC8 is equal to the reference frequency ÷ by 1024. Thus, the output from pin 7 is 5.12 MHz ÷ 1024, or 5 kHz. This 5-kHz signal is applied to pin 7 of IC9, where it serves as the reference signal. The other input to IC9 comes from the output of the

Table 11-8. Various Pin Functions of the TC5082P.

PIN NO.	PIN ASSIGNMENT
1	This is a buffered output of the reference oscillator signal. Usually 10.240 MHz. Symbol = XTout
2	This is the input to the on-chip oscillator used for the reference frequency generator (with an external xtal). Symbol = XT
3	This is the output of the on-chip oscillator. Symbol = XT
4	This output at this terminal is equal to the reference frequency divided by 4096. Symbole = Q12
5	This is the supply voltage input to the IC. Symbol = VDD
6	The output at this terminal is equal to the reference frequency divided by 2048. Symbol = Q11
7	The output at this terminal is equal to the reference frequency divided by 1024. Symbol = Q10
8	The output at this terminal is equal to the reference frequency divided by 256. Symbol = Q8
9	This is the ground terminal of the IC. Symbol = VSS

Table 11-9. Various Divide Ratios of the TC5082P.

PIN NUMBER	8	7	6	4	1
SYMBOL	Q8	Q10	Q11	Q12	XTout
DIVIDE RATIO	1/256	1/1024	1/2048	1/4096	1/1
10.240 MHz Ref.	40 kHz	10 kHz	5 kHz	2.5 kHz	10.24 MHz

programmable divider, IC10. The control voltage at the output of IC9 is applied to the control line of the voltage-controlled multivibrator. The voltage-controlled multivibrator consists of transistors Q40 and Q39. This is a stable emitter-coupled multivibrator. Notice that the emitter/collector of Q41 is connected in the emitter circuit of Q40 and the emitter/collector of Q42 is connected in the emitter circuit of Q39. The running frequency of this multivibrator is determined by capacitor C191 and the resistance between Q39 and ground and the resistance between the emitter of Q40 and ground. Since transistors Q41 and Q42 serve as part of the emitter resistances, the running frequency of the multivibrator can be controlled by controlling the bias to transistors Q41 and Q42.

The multivibrator output is fed to the rf amplifier. The rf amplifier amplifies this signal and feeds it to the input of the programmable divider. Also, part of the multivibrator signal is fed from the rf amplifier to the reference input of phase comparator, IC7. Thus, the multivibrator serves as the reference frequency generator for phase comparator IC7. The frequency at the output of programmable divider IC10 should always be 5 kHz because it

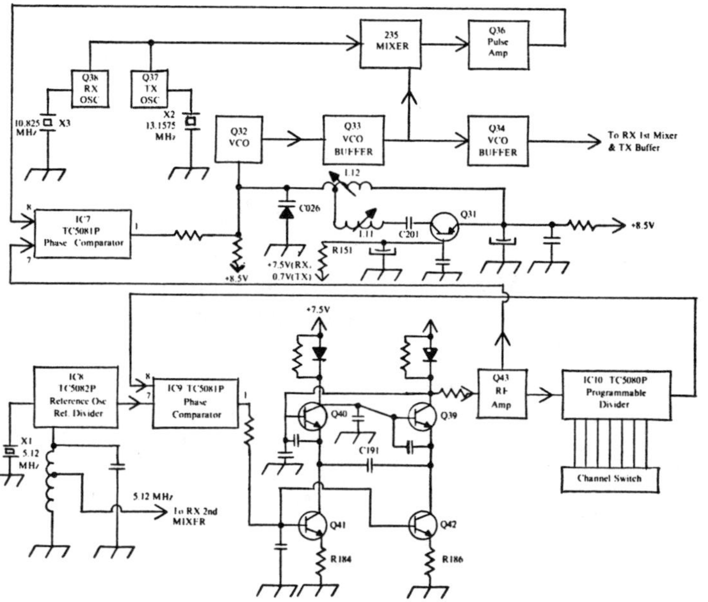

Fig. 11-10. Partial schematic and block diagram of the TC5082P (courtesy of Toshiba).

feeds the input of phase comparator IC9, which has a reference frequency of 5 kHz. In order for the output of the programmable divider to be 5 kHz, the multivibrator frequency will have to be fM = N × 5 kHz.

On CB channel 1, the N value programmed into the programmable divider is 130. This sets the multivibrator frequency at fM = 130 × 5 kHz, or 650 kHz or 0.65 MHz. If the multivibrator IC9 will be greater than 5 kHz. This will cause a decrease in the control voltage that is applied to the base of the control transistors, Q41 and Q42. This decrease in bias voltage will increase the effective emitter/collector resistance of the two control transistors. This will cause the multivibrator to run at a lower frequency. If the multivibrator frequency is lower than 0.65 MHz, just the opposite situation occurs. Thus, the multivibrator is kept on the proper frequency by the action of the phase comparator.

The 0.65-MHz multivibrator signal is also fed from rf amplifier Q43 to the reference input of phase comparator IC7. The other input to phase comparator IC7 comes from mixer Q35 by way of pulse amplifier Q36. In the receive mode, receive oscillator

Q38 generates a frequency of 10.825 MHz. The VCO frequency is heterodyned with the second harmonic of 10.825 MHz (21.65 MHz) to produce a difference frequency at the output of the mixer which is equal to the reference input of phase comparator IC7.

In the receive mode, the VCO operates in a frequency range from 22.300 MHz to 22.740 MHz. In the transmit mode, the VCO operates in a frequency range from 26.965 MHz to 27.405 MHz. The VCO frequency for the receive mode can be determined from the following formula: fVCO = fM + 21.65 MHz, where fVCO is the VCO frequency in MHz, and fM is the multivibrator frequency in MHz. The figure -21.65 MHz is the second harmonic of the 10.825-MHz receive oscillator. Since the multivibrator frequency for channel 1 is 0.65 MHz, you can substitute this into the formula to find the receive mode VCO frequency. Substituting, you have fVCO = 0.65 MHz + 21.65 MHz, or 22.300 MHz. This 22.300-MHz signal is fed to the receive first mixer. The output circuit of the first mixer is tuned to 4.665 MHz, which is the difference between the VCO frequency and 26.965 MHz, the CB channel 1 frequency. This 4.665-MHz is the first receive i-f. This 4.665-MHz first i-f signal is then mixed with a 5.12-MHz signal from the reference oscillator. The difference of these two frequencies is 5.12 MHz - 4.665 MHz, or 455 kHz. This is the receive second intermediate frequency.

In the transmit mode, the receive oscillator (10.825 MHz) is switched out of the circuit and the 13.1575-MHz transmit oscillator is switched into the circuit. This will force the VCO to operate at a much higher frequency than it did in the receive mode. The transmit mode VCO frequency will have to be 4.665 MHz higher than the receive mode VCO frequency. However, the output of the phase detector can't affect such a large change in VCO frequency so a special circuit is used to increase the VCO frequency range in the transmit mode. In the receive mode, transistor Q31 is cut off so coil L11 is effectively disconnected from the circuit. In the transmit mode, Q31 conducts, thus connecting L11 in parallel with part of L12. When two inductors are connected in parallel, the total inductance of the combination is determined by the formula: L(total) = L1 × L2/ L1 + L2. So when L11 is connected into the circuit, the total inductance is lowered. And when the inductance of a resonant circuit is lowered, the resonant frequency increases. The VCo frequency for the transmit mode is determined by the following formula: fVCO = fM + 26.315 MHz. For channel 1 transmit, the VCO frequency is: fVCO = 0.65 MHz + 26.315

MHz, or 26.965 MHz. Thus, the VCO operates at the correct transmit frequency for each channel. No further mixing of the VCO frequency is required.

Table 11-10 lists the transmit and receive mode VCO frequencies, the multivibrator frequencies, the programming code, and the N values for each of the 40 channels. The reason that the N value increases in increments of 2 is because the reference frequency of the phase comparator, IC9 is 5 kHz.

Here is a brief summary of the operation of this PLL circuit. The phase comparator, IC9, forces the voltage-controlled multivibrator to run at a frequency, which when divided by N, will be equal to its reference frequency of 5 kHz. The multivibrator operates at a different frequency for each channel because of the changing N values of the programmable divider. The multivibrator frequency serves as a reference frequency for phase comparator IC7. The output of IC7 forces the VCO to operate at a frequency that keeps the frequency at pin 8 of IC7 equal to the reference frequency at pin 7. For instance, on channel 1 the reference frequency at pin 7 of IC7 will be 0.65 MHz. The VCO will be forced to run at 22.3 MHz in the receive mode. This 22.3-MHz VCO frequency mixes with 21.65 MHz (second harmonic of the receive oscillator) in the mixer to produce a difference output of 0.65 MHz. This 0.65-MHz signal is then fed to the pulse amp (Q36) and on to the input of IC7 at pin 8.

Most of the PLL circuits you will encounter are not so complex. Compare this circuit with the one used in the Sharp model CB-2260, which is described next.

SHARP MODEL CB-2260 PLL SYNTHESIZER

The complete schematic diagram of the Sharp model CB-2260 PLL synthesizer circuit is shown in Fig. 11-11. A brief description of the operation of each stage of this circuit follows.

Programmable Divider

First, notice that pin 11 of IC203 is grounded. This means that the programming to the program lines will be done using the pure binary code. Table 11-11 lists the programming code for each channel. Above the program pin number is the weight of the pin. To get the N value, simply add the weights where the binary 1 appears. Two of the program pins (7 and 8) are not connected to the channel switch. Instead, pin 7 is grounded so its weight of 64 will *never* be used in determining the divisor N. Pin 8 is connected directly to binary 1 (6.4 volts) so its weight of 128 is *always* used to

Table 11-10. Tram D12 CB PLL Circuit Specifications.

CH. No.	N	Pin No. Weight	8 128	7 64	6 32	5 16	4 8	3 4	2 2	Rx VCO Freq.	Tx VCO Freq.	Multi-vibrator Freq.
1	130		1	0	0	0	0	0	1	22.300	26.965	.650
2	132		1	0	0	0	0	1	0	22.310	26.975	.660
3	134		1	0	0	0	0	1	1	22.320	26.985	.670
4	138		1	0	0	0	1	0	1	22.340	27.005	.690
5	140		1	0	0	0	1	1	0	22.350	27.015	.700
6	142		1	0	0	0	1	1	1	22.360	27.025	.710
7	144		1	0	0	1	0	0	0	22.370	27.035	.720
8	148		1	0	0	1	0	1	0	22.390	27.055	.740
9	150		1	0	0	1	0	1	1	22.400	27.065	.750
10	152		1	0	0	1	1	0	0	22.410	27.075	.760
11	154		1	0	0	1	1	0	1	22.420	27.085	.770
12	158		1	0	0	1	1	1	1	22.440	27.105	.790
13	160		1	0	1	0	0	0	0	22.450	27.115	.800
14	162		1	0	1	0	0	0	1	22.460	27.125	.810
15	164		1	0	1	0	0	1	0	22.470	27.135	.820
16	168		1	0	1	0	1	0	0	22.490	27.155	.840
17	170		1	0	1	0	1	0	1	22.500	27.165	.850
18	172		1	0	1	0	1	1	0	22.510	27.175	.860
19	174		1	0	1	0	1	1	1	22.520	27.185	.870
20	178		1	0	1	1	0	0	1	22.540	27.205	.890
21	180		1	0	1	1	0	1	0	22.550	27.215	.900
22	182		1	0	1	1	0	1	1	22.560	27.225	.910
23	188		1	0	1	1	1	1	0	22.590	27.255	.940
24	184		1	0	1	1	1	0	0	22.570	27.235	.920
25	186		1	0	1	1	1	0	1	22.580	27.245	.930
26	190		1	0	1	1	1	1	1	22.600	27.265	.950
27	192		1	1	0	0	0	0	0	22.610	27.275	.960
28	194		1	1	0	0	0	0	1	22.620	27.285	.970
29	196		1	1	0	0	0	1	0	22.630	27.295	.980
30	198		1	1	0	0	0	1	1	22.640	27.305	.990
31	200		1	1	0	0	1	0	0	22.650	27.315	1.000
32	202		1	1	0	0	1	0	1	22.660	27.325	1.010
33	204		1	1	0	0	1	1	0	22.670	27.335	1.020
34	206		1	1	0	0	1	1	1	22.680	27.345	1.030
35	208		1	1	0	1	0	0	0	22.690	27.355	1.040
36	210		1	1	0	1	0	0	1	22.700	27.365	1.050
37	212		1	1	0	1	0	1	0	22.710	27.375	1.060
38	214		1	1	0	1	0	1	1	22.720	27.385	1.070
39	216		1	1	0	1	1	0	0	22.730	27.395	1.080
40	218		1	1	0	1	1	0	1	22.740	27.405	1.090

Table 11-11. Sharp CB-2260 CB PLL Circuit Specifications.

CH.	N	Program pin Weight	6 32	5 16	4 8	3 4	2 2	1 1	VCO Freq. MHz	Offset Freq. MHz
1	128		0	0	0	0	0	0	38.240	1.28
2	129		0	0	0	0	0	1	38.250	1.29
3	130		0	0	0	0	1	0	38.260	1.30
4	132		0	0	0	1	0	0	38.280	1.32
5	133		0	0	0	1	0	1	38.290	1.33
6	134		0	0	0	1	1	0	38.300	1.34
7	135		0	0	0	1	1	1	38.310	1.35
8	137		0	0	1	0	0	1	38.330	1.37
9	138		0	0	1	0	1	0	38.340	1.38
10	139		0	0	1	0	1	1	38.350	1.39
11	140		0	0	1	1	0	0	38.360	1.40
12	142		0	0	1	1	1	0	38.380	1.42
13	143		0	0	1	1	1	1	38.390	1.43
14	144		0	1	0	0	0	0	38.400	1.44
15	145		0	1	0	0	0	1	38.410	1.45
16	147		0	1	0	0	1	1	38.430	1.47
17	148		0	1	0	1	0	0	38.440	1.48
18	149		0	1	0	1	0	1	38.450	1.49
19	150		0	1	0	1	1	0	38.460	1.50
20	152		0	1	1	0	0	0	38.480	1.52
21	153		0	1	1	0	0	1	38.490	1.53
22	154		0	1	1	0	1	0	38.500	1.54
23	157		0	1	1	1	0	1	38.530	1.57
24	155		0	1	1	0	1	1	38.510	1.55
25	156		0	1	1	1	0	0	38.520	1.56
26	158		0	1	1	1	1	0	38.540	1.58
27	159		0	1	1	1	1	1	38.550	1.59
28	160		1	0	0	0	0	0	38.560	1.60
29	161		1	0	0	0	0	1	38.570	1.61
30	162		1	0	0	0	1	0	38.580	1.62
31	163		1	0	0	0	1	1	38.590	1.63
32	164		1	0	0	1	0	0	38.600	1.64
33	165		1	0	0	1	0	1	38.610	1.65
34	166		1	0	0	1	1	0	38.620	1.66
35	167		1	0	0	1	1	1	38.630	1.67
36	168		1	0	1	0	0	0	38.640	1.68
37	169		1	0	1	0	0	1	38.650	1.69
38	170		1	0	1	0	1	0	38.660	1.70
39	171		1	0	1	0	1	1	38.670	1.71
40	172		1	0	1	1	0	0	38.680	1.72

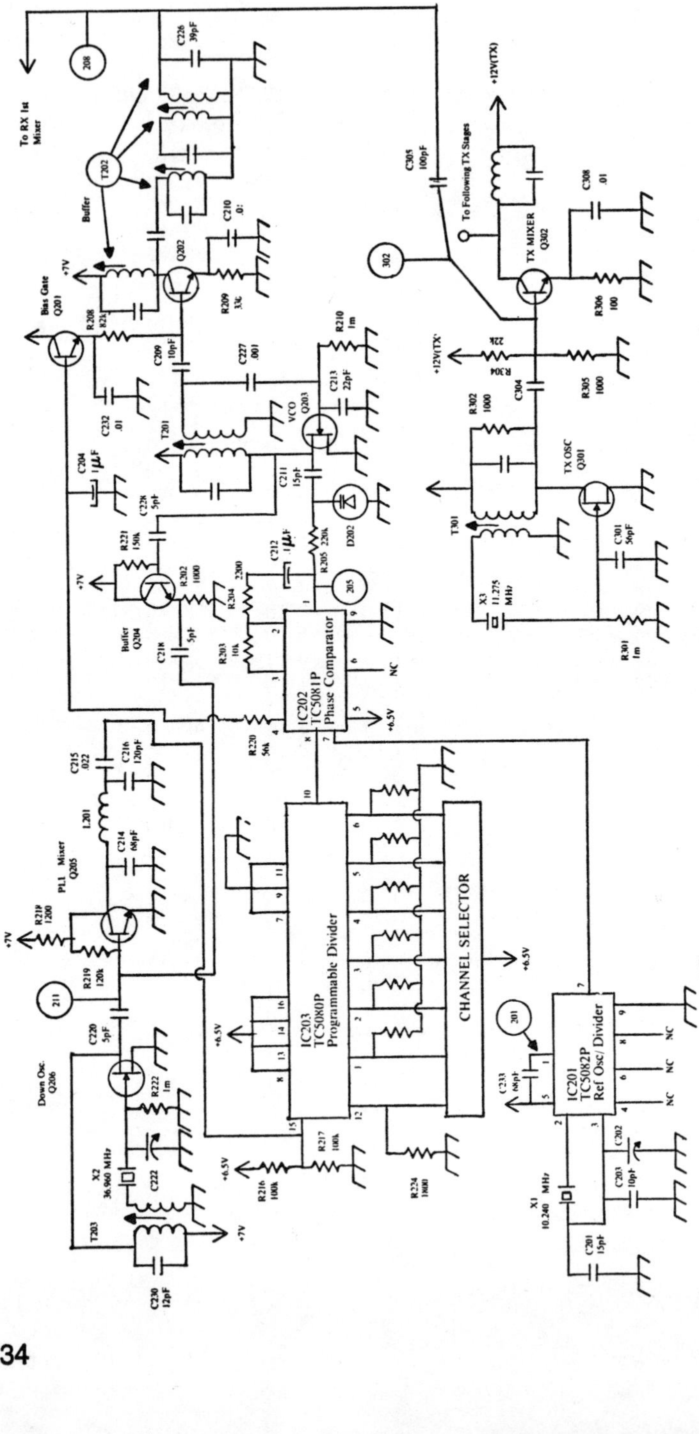

Fig. 11-11. Schematic diagram of the PLL circuit used in the Sharp CB2260 CB radio set.

determine the N value. Notice from Table 11-11 that on channel 1, programming pins 1 through 6 are at binary 0. Thus, the N value will be 128 because only pin 8 is at the binary 1 level.

External pulldown resistors are used on the program lines. These pulldown resistors are 56K ohm resistors, all housed in a single package. The channel selector switch places binary 1 level voltage on the appropriate pins for each channel. The pins which are left open by the channel selector switch are pulled down to binary 0 level by the pulldown resistors.

You may have wondered why pin 12 of IC203 is connected to the channel select switch since this is not a program input. Pin 12 is the RESET terminal. If the channel selector switch is set between channels, a HIGH level voltage will be applied to pin 12 from the channel selector. This HIGH level voltage causes all stages of the counter to clear. Thus there will be no output from pin 10 during this condition. With no output from the programmable divider, the phase comparator (IC202) lock detector output at pin 4 will go LOW. This kills the forward bias to the VCO buffer-Q202, thus stopping the transmitter output. Under normal conditions (when the channel selector is properly positioned), pin 12 of IC203 is pulled down to a LOW level by R224. This enables the counter to work.

The input to the programmable divider at pin 15 is taken from the output of the PLL mixer stage. Positive bias to pin 15 is supplied from the junction of R216 and R217. The output from pin 10 of the programmable divider is fed to the input of the phase comparator-IC202.

Phase Comparator

The 10-kHz reference frequency for the phase detector is applied to pin 7. The frequency to be checked or compared is applied to pin 8. Pin 3 of the IC is the output of the charge pump. This is fed through resistor R203 to pin 2, which is the input to the on-chip filter amplifier. The output of the filter amplifier is then taken from pin 1 and applied to the VCO control line. An RC network consisting of R204 and C212 is connected between the input and output of the on-chip filter amplifier. This determines the characteristics of the filter amplifier.

VCO and VCO Buffers

The VCO operates in the 38-MHz range. The VCO oscillator is built around FET transistor Q203. The source is grounded. The resonant circuit for this oscillator is connected in the drain circuit. This resonant circuit consists of the primary of T201, capacitors C208 and C211 and varactor diode D202. Feedback from the oscillator is furnished by capacitor C227, which feeds part of the signal from the secondary of T201 back to the gate of Q203. Notice that the cathode of the varactor is connected to the control line from the phase comparator. If the positive voltage on the control line increases, the capacitance of the varactor decreases, thus increasing the VCO frequency. Conversely, if the positive voltage on the control line decreases, the capacitance of the varactor increases, thus decreasing the VCO frequency.

One output of the VCO is taken from the secondary of T201 and applied to the base of Q202. Transistor Q202 amplifies this VCO signal and passes it on to the receive first mixer and to the transmit mixer stage. Another output from the VCO is taken from the drain of Q203 and fed to the base of Q204. Transistor Q204 is connected as an emitter follower, so no voltage gain is achieved but buffering is accomplished so that the VCO is not loaded down. The buffered VCO signal is fed from the emitter of Q204 to the PLL mixer stage.

Bias Gate

The function of the bias gate is to remove bias from the VCO buffer, Q202, when the loop is unlocked. Normally the voltage at pin 4 of IC202 is a HIGH level. This HIGH level voltage is used to forward bias Q201 into conduction. With Q201 conducting, forward bias is supplied to the base of Q202 through Q201. If the loop becomes unlocked, the voltage at pin 4 of IC202 will go LOW. This will cause Q201 (the bias gate) to turn off, thus killing the forward bias to VCO buffer Q202. This stops the transmitter output.

Down Oscillator

The function of the down oscillator is to supply a stable reference signal to beat against the VCO frequency to produce a difference frequency which is low enough to be handled by the programmable divider. This down oscillator operates at a frequency of 36.960 MHz. Crystal X2 controls this frequency very closely. The frequency can be adjusted by capacitor C222. Feedback for oscillation is supplied through transformer T203, which feeds the signal from the drain circuit back to the input (gate circuit). The output of the oscillator is then taken from the drain through capacitor C220.

PLL Mixer

The PLL mixer is a simple common-emitter amplifier. The base is fed two input signals, one from the 36.960-MHz down oscillator and the other from the VCO buffer, Q204. These two signals are heterodyned in the mixer. The output of the mixer is tuned to the difference of these two frequencies by the low-pass filter. The low-pass filter consists of L201, C214, and C216. While the other frequencies are attenuated by the filter, the difference frequency passes through the filter on to the input of the programmable divider.

Transmit Oscillator

The transmit oscillator operates at 11.275 MHz. This frequency is used to heterodyne with the VCO frequency to yield a difference frequency which is the transmit frequency. The primary of transformer T301 is connected in the drain circuit of Q301. The signal developed in the secondary of T301 is fed back to the gate of Q301 through crystal X3. The frequency of the oscillator can be adjusted by tuning T301. The output is taken from the collector.

Transmit Mixer

The transmit mixer is a common-emitter amplifier which has two signals applied to its input. One of the signals is from the 11.275-MHz transmit oscillator. The other is the VCO signal which comes from VCO buffer Q202. These two signals are heterodyned to produce a difference frequency at the output that is equal to the transmit frequency of the particular CB channel. The other frequencies generated by the mixing process are attenuated by the tuned circuit in the collector circuit of Q302. The output of the mixer is fed on to the following transmit stages.

Circuit Analysis

When the radio is switched to channel 40, the N value programmed into the programmable divider will be 172. Referring to Table 11-11, add the weights of the program pins where binary 1 appears. This is 32 + 8 + 4, or 44. Now add 128 to 44 because pin 8 is always at binary 1. This gives an N value of 128 + 44, or 172. Since the reference frequency for the phase comparator at pin 7 of IC202 is 10 kHz, the phase comparator has to force the VCO to run at a frequency that will provide a 10-kHz signal on pin 8 of IC202 to match the reference frequency. Thus, the output of the programmable divider must be kept at 10 kHz. For channel 40, where the N divisor is 172, the input to the programmable divider will have to be 172 × 10 kHz, or 1.72 MHz. In order for the frequency at the programmable divider input frequency to be 1.72 MHz, the VCO frequency must be 1.72 MHz higher than the 36.960-MHz down oscillator frequency. This would make the VCO frequency 36.960 MHz + 1.72 MHz, or 38.680 MHz. the action of the phase comparator holds the VCO at that frequency.

In the receive mode, the 38.680-MHz VCO signal is fed to the receive first mixer. The first intermediate frequency has been chosen as 11.275 MHz so the output circuit of the first receive mixer is sharply tuned to this frequency. The receive frequency in this case will be 38.680 MHz - 11.275 MHz, or 27.405 MHz, which is the correct channel 40 frequency. This 11.275-MHz first i-f signal is then mixed with a 11.730- MHz second oscillator signal to produce a second intermediate frequency of 11.730 MHz -11.275 MHz, or 455 kHz. The second oscillator and mixer are not shown in the schematic.

In the transmit mode, the VCO signal is fed to the base of transmit mixer Q302, along with the 11.275-MHz transmit oscillator signal. These two frequencies are mixed to yield a difference frequency of 38.680 MHz - 11.275 MHz, or 27.405 MHz. This is the correct channel 40 transmitter frequency.

Alignment and Troubleshooting

□ Connect a frequency counter to pin 1 of IC201 (TP201) and adjust C202 for exactly 10.240 MHz.

□ Connect an rf voltmeter to the base of Q205 (TP211) and adjust T203 for maximum rf voltage. Then detune T203 just slightly so that it is below the peak on the gentle slope of the response curve. This improves the stability of the oscillator and prevents erratic operation.

□ Connect a frequency counter to the base of Q205 and adjust C222 for exactly 36.960 MHz. To prevent interference from the VCO signal, you can temporarily connect a 0.01-muF capacitor between the base of Q204 and ground.

□ Set the channel selector to channel 20 and adjust T201 for exactly 3.0 volts DC at pin 1 of IC202 (TP205).

□ Connect a frequency counter to the base of Q302 (TP302). Key the transmitter and adjust T301 for exactly 11.275 MHz. To

Table 11-12. Troubleshooting Chart.

SYMPTOM	POSSIBLE CAUSE
No Tx or Rx any channel Loop Unlocked	Any stage associated with the loop could cause this trouble. Check following: Reference osc/div stage, Programmable divider stage, Phase comparator VCO, VCO buffer-Q204, down oscillator, PLL mixer
No Tx or Rx any channel Loop Locked	Probably bias gate stage or VCO buffer-Q202
No Tx, Rx OK all channels	Tx oscillator-Q301, Tx miser-Q302
Tx & Rx off frequency all channels	Reference oscillator, (possibly) Down oscillator, (more likely)
Tx off frequency, Rx OK	Tx oscillator off frequency
High channels inoperative Low channels OK	VCO misaligned
Low channels inoperative High channels OK	VCO misaligned
Some channels inoperative (Scattered)	Programmable divider or program lines bad
Some channels intermittent	Channel selector switch

prevent interference from the VCO signal, temporarily connect a 0.01 muF capacitor between the base of Q202 and ground.

☐ Connect a frequency counter to TP208 and check for the proper VCO frequency for each channel. Refer to Table 11-11 for the proper VCO frequency for each channel.

☐ Connect an rf voltmeter to TP208 and adjust T202 for maximum rf voltage.

☐ Refer to Table 11-12 for troubleshooting tips.

Appendix A
The PLL03A PLL IC

The PLL03A is a shift-type, CMOS PLL IC device in a 16-pin dual-in-line plastic package. The top view of the PLL03A is shown in Fig. A-1. The functional block diagram of the IC is shown in Fig. A-2. Notice in Fig. A-2 that this IC has an internal code converter between the program input lines and the programmable divider. The programming code is applied to the code converter, which then applies another code to the programmable divider for a particular N number.

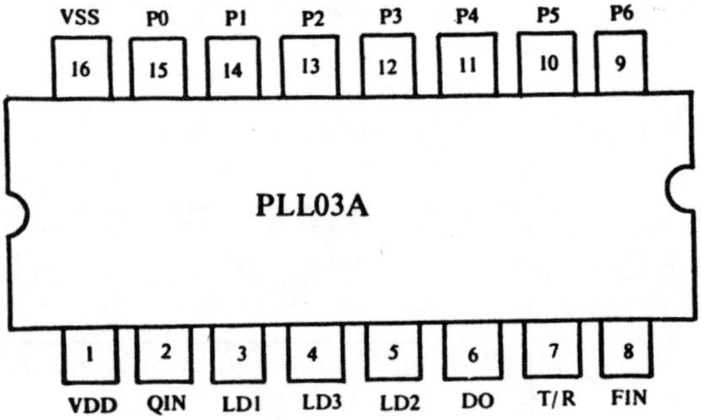

Fig. A-1. Top view of the PLL03A PLL IC.

PIN FUNCTIONS

☐ Pin 1. This is the supply voltage terminal. This IC requires a nominal VDD supply of + 6 volts.

☐ Pin 2. This is the input to the reference divider. Normally the reference frequency at this terminal is 10.240 MHz.

☐ Pin 3. This is connected to the input of an inverter stage. Normally this terminal is connected to the channel selector switch. If the channel selector switch is improperly positioned between channels, a HIGH level will be applied to LD1.

☐ Pin 4. This is the lock detector output of the phase detector. When the loop is locked, this terminal will be at a HIGH level. When the loop is unlocked, this terminal goes to a LOW level. Notice that the lock detector output is connected to the gate of an FET, which is connected as a source follower. The source con-

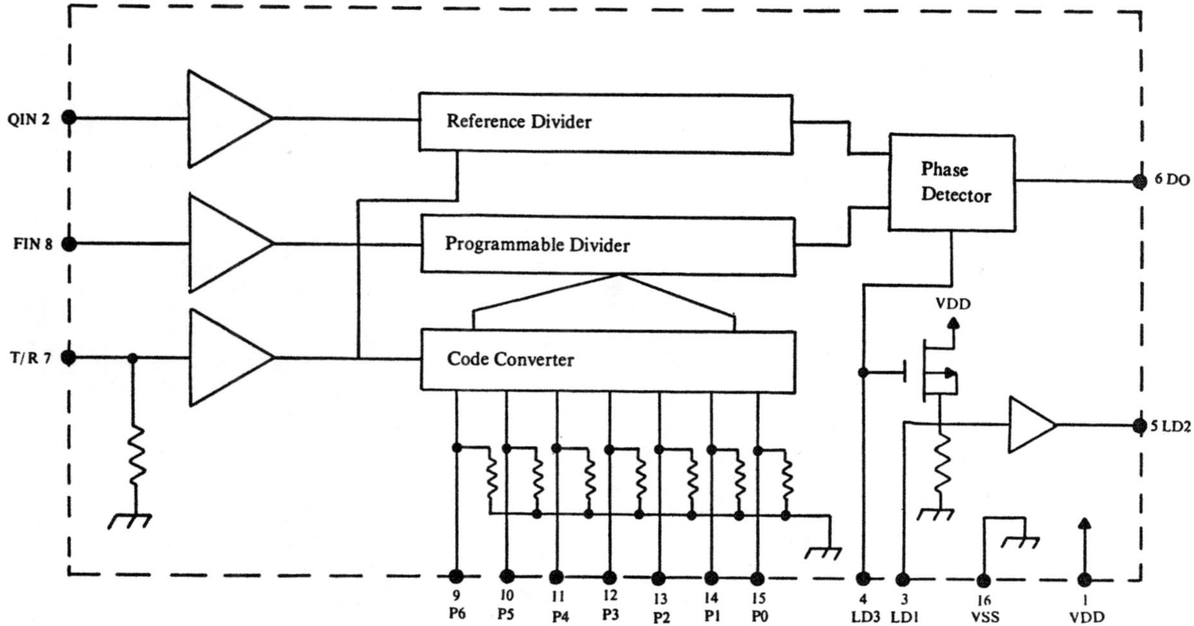

Fig. A-2. Functional block diagram of the PLL03A. Notice the code converter between the program inputs and the programmable divider. This converts the input code to a specific N divisor for the receive mode and to N + 91 in the transmit mode.

nects to an inverter input same as pin 3.

□ Pin 5. This is the output of an inverter. If the lock detector output (at pin 4, LD3) is HIGH and pin 3 (LD1) is open, the output of LD2 will be HIGH. If the lock detector output (LD3) goes LOW or if the LD1 terminal (pin 3) goes HIGH, the output of LD2 (pin 5) goes LOW.

□ Pin 6. This is the phase detector output. This is a three-state output. If F-IN is too low as compared to Q-IN, the DO terminal is at a HIGH state. If F-IN is too HIGH as compared to Q-IN, then DO is at a LOW state. If F-IN is at the proper frequency, the DO terminal goes to an open or high-impedance state.

□ Pin 7. This terminal is used to change the divide ratio of both the programmable divider and the reference divider. When this terminal is LOW or open, the reference divider ratio is 2048 and the programmable divider ratio is N, as determined by the program code. When the T/R terminal is HIGH, the reference divider ratio becomes 4096 and the programmable divider ratio becomes N + 91.

□ Pin 8. This is the input to the programmable divider. The maximum rated frequency at this input is 7 MHz.

□ Pin 9 through 15. These are the programming inputs to the IC.

□ Pin 16. This is the ground terminal for the IC.

Appendix B
The M58473P PLL IC

The M58473P is an 18-pin dual-in-line PLL IC in a plastic package. The device is manufactured by Mitsubishi Electric Corp. (MELCO). A top view of the IC is shown at Fig. B-1. Specifications are lists in Tables B-1 and B-2. The functional block diagram of the M58473P is shown at Fig. B-2. Notice in Fig. B-2 that there is a code converter connected between programming input lines (D1 through D6) and the programmable divider. Table B-3 shows how the programming code relates to the divide ratio of the programmable divider. In Table B-3, notice that when all programming inputs are at binary 0 *and* the EX terminal (pin 14) is at binary 1 (open), the programmable divider divide ratio will be 147. If the EX terminal is grounded, the divide ratio will be 203. If the programmable divider ratio is defined as N when EX is binary 1 (open), the divide ratio becomes 2N - 91 when the EX terminal is binary 0 (grounded).

The phase detector reference frequency is also affected by the binary level at the EX terminal. When the EX terminal is binary 1, the input reference oscillator frequency is divided by 1024. When the EX terminal is binary 0 (grounded), the input reference oscillator frequency is divided by 2048. Pullup resistors are installed on the programming lines, D1 through D6, and on the EX terminal.

If this IC is used in a two-crystal PLL circuit, the EX terminal will be used. In a three-crystal PLL circuit, where a separate transmit oscillator is used, the EX terminal will not be used.

PIN FUNCTIONS

☐ Pin 1 through 3. Program input lines.

☐ Pin 4. Clock input of the programmable divider.

☐ Pin 5. Clock input of the reference dividers or reference oscillator input.

☐ Pin 6. This is the output of the on-chip amplifier which feeds the input to the programmable divider. If the on-chip amplifier is used as the reference oscillator, this is the OSC/OUT terminal.

☐ Pin 7. This is a buffered output of the reference dividers.

☐ Pin 8. This is the unlock terminal. If the loop is locked,

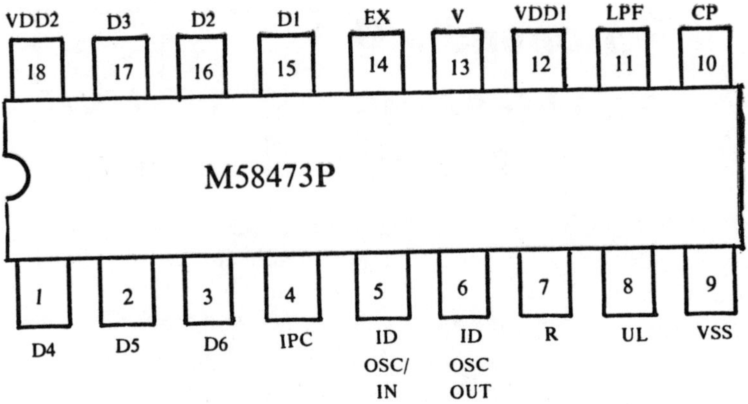

Fig. B-1. Top view of the M58473P PLL IC.

Table B-1. Maximum Ratings of the M58473P (courtesy of Mitsubishi Electric Corp.).

ABSOLUTE MAXIMUM RATINGS:

Power Supply Voltage	(VDD - VSS)	-0.3 to +9.0 V
All inputs	(VI)	VSS = VI = VDD
Operating Temperature Range	(Topg)	-30 to +70° C
Storage Temperature Rage	(Tstg)	-40 to +125° C

Table B-2. Specifications of the M58473P (courtesy of Mitsubishi Electric Corp.).

OPERATING CONDITIONS AT TA=25° C:

Descriptions	Symbol	Test Conditions	LIMITS Min	Typ	Mx	Unit
Operating Voltage	VDD-VSS		6.0	7.0	8.0	V
Supply Current	IDD	VDD=7.0 V fpc=2.1 MHz fx=10.24 MHz	-	1.5	5	mA
Maximum frequency of Programmable Counter	fPC MAX	VDD=7.0 V	3.5	-	-	MHz
Output drive current (UL Output)	IOH	VDD=7.0 V RL=1k to VSS	1	-	-	mA
	IOL	VDD=7.0 V RL=1k (to VDD)	1	-	-	mA
Low Input Current (D1 D6, Ex Inputs)	IIL	VDD=7.0 V VIN=O V	25	70	250	uA

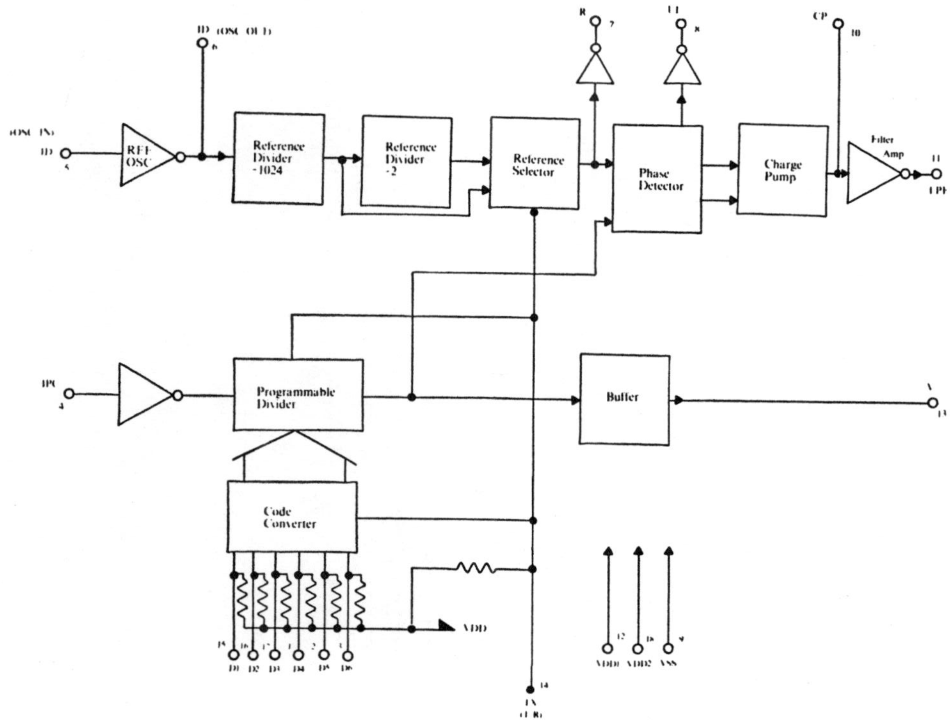

Fig. B-2. Functional block diagram of the M58473P (courtesy of Mitsubishi Electric Corp.).

Table B-3. Programming Code Versus Divide Ratio (courtesy of Mitsubishi Electric Corp.).

Programming Code						Divide Ratio	
D1	D2	D3	D4	D5	D6	(N) EX = 1	(2N-91) EX = 0
0	0	0	0	0	0	147	203
1	0	0	0	0	0	148	205
0	1	0	0	0	0	149	207
0	1	1	1	1	1	209	327
1	1	1	1	1	1	210	329

the voltage at this pin is LOW. If the loop is unlocked, the voltage at this pin is HIGH.

☐ Pin 9. This is the ground terminal of the IC.

☐ Pin 11. This is the output of the on-chip fiter amplifier-low pass filter.

☐ Pin 12. This is a voltage supply point for the IC.

☐ Pin 13. This is the output of the programmable divider.

☐ Pin 14. This terminal is used between the transmit and receive modes in some PLL circuits to change the divide ratio of the programmable divider and the reference divider. When EX is open or HIGH, the programmable divider divides by N and the reference frequency divider divides by 1024. When EX is grounded, the programmable divider divides by 2N - 91 and the reference divider divides by 2048.

☐ Pins 15 through 17. Program input lines.

☐ Pin 18. This is a voltage supply point for the IC.

Appendix C

The MSM5807 PLL IC

The MSM5807 IC is a CMOS 16-pin dual-in-line IC in a plastic package. An outline drawing of the MSM5807 is shown in Fig. C-1. The nominal operating voltage (VDD) is 6 volts. The maximum clock frequency of the programmable divider is 5 MHz. The maximum divisor of the programmable divider (N) is 255. There is no code converted used in this IC so the N value will be equal to the decimal equivalent of the programming code.

The internal structure of the MSM5807 is shown in the functional block diagram in Fig. C-2. There is an on-chip amplifier which can serve as the reference oscillator by connecting a bias resistor and crystal between pins 6 and 7. There is also a reference frequency selector switch which determines whether the reference oscillator frequency is divided by 512 or 1024. Notice also that each of the program pins (DP1 through DP8) have an on-chip pulldown resistor on them. Program line DP1 is the least significant bit (1) while program line DP8 is the most significant bit (128).

PIN FUNCTIONS

☐ Pins 1 through 4. These are programming inputs to the programmable divider.

☐ Pin 5. When this pin is at a HIGH level (binary 1), the reference dividers divide by 512. When this pin is at a LOW level (binary 0) or ground, the reference dividers divide by 1024.

☐ Pin 6. This is the input side of the on-chip amplifier.

☐ Pin 7. This is the output of the on-chip amplifier.

☐ Pin 8. This is the ground terminal for the IC.

☐ Pin 9. This is the three-state output of the phase comparator.

☐ Pin 10. This is the output of the lock detector. During locked conditions, the LD terminal is at a LOW level. When the loop is unlocked, the LD terminal goes to a HIGH level (VDD).

☐ Pin 11. This is the input to the programmable divider.

☐ Pins 12 through 15. These are the programming inputs.

☐ Pin 16. This is the voltage supply point for the IC.

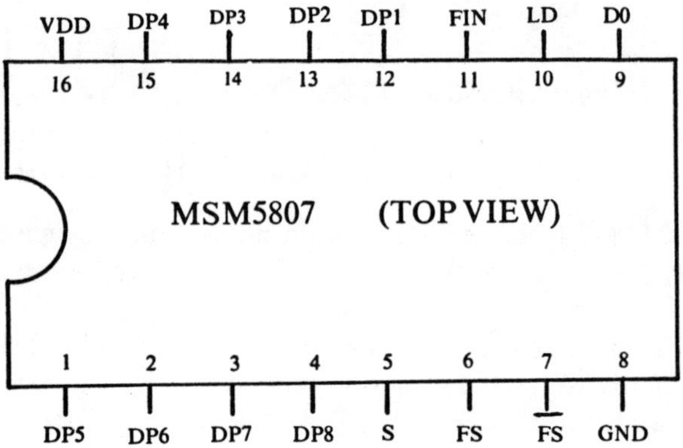

Fig. C-1. Top view of the MNM5807 PLL IC (courtesy of OKI Semiconductor).

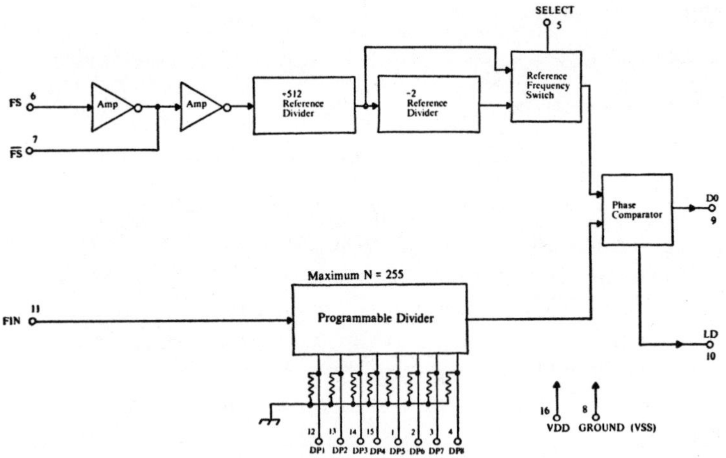

Fig. C-2. Functional block diagram of the MSM5807.

Appendix D National Semiconductor PLL ICs

The following information is reprinted by permission of National Semiconductor Corp.

MM55104, MM55106, MM55107, MM55114, MM55116 PLL FREQUENCY SYNTHESIZER

The MM55104, MM55106, MM55107, MM55114 and MM55116 devices contain phase locked loop circuits useful for frequency synthesizer applications in CB transceivers. The devices operate from a single power supply and contain an oscillator, a 2^{10} or 2^{11} divider chain, a binary input programmable divider, and phase detectir circuitry. The devices may be used in double IF or single IF systems. The MM55104, MM55106, MM55107, MM55114 and MM55116 use a 10.24 MHz or 5.12 MHz quartz crystal to determine the reference frequency. The MM55106, MM55107 and MM55116 have an output pin which provides a 5.12 MHz signal, which may be tripled for use as a reference oscillator frequency in two crystal systems. Also, the MM55106 and MM55116 provide an additional input to the programmable divider which allows $2^9 - 1$ division of the input frequency (FIN). The inputs to the programmable divider are standard binary signals. Selection of a channel is accomplished by mechanical switches or by external electronic programming of the programmable divider.

The øVCO output provides a high level voltage (sources current) when the VCO frequency is lower than the lock frequency, and øVCO provides a low level voltage (sinks current) when the VCO frequency is higher than the lock frequency. The øVCO output goes to a high impedance (TRISTATE®) condition under lock conditions, and the lock detector output LD goes to a high state under lock conditions.

Features

- ☐ Single power supply
- ☐ Low power CMOS technology
- ☐ Binary input channel select code
- ☐ 5 kHz or 10 kHz output from oscillator divide

□ 5.12 MHz output (MM55106, MM55107 and MM55116 only)

□ On-chip oscillator

□ Pulldown resistors on programmable divider inputs

□ Low voltage operation—5V (MM55104, MM55106 and MM55107)

□ High voltage operation - 8V (MM55114, MM55116)

Introduction To Frequency Synthesis

The components of a frequency synthesizer are shown in Fig. D-1. The voltage controlled oscillator produces the desired output frequencies spaced fv Hz apart according to the relation:

$$fv = fr\ N$$

The reference frequency, fr, must be equal to or less than the (channel) spacing between the frequencies being synthesized.

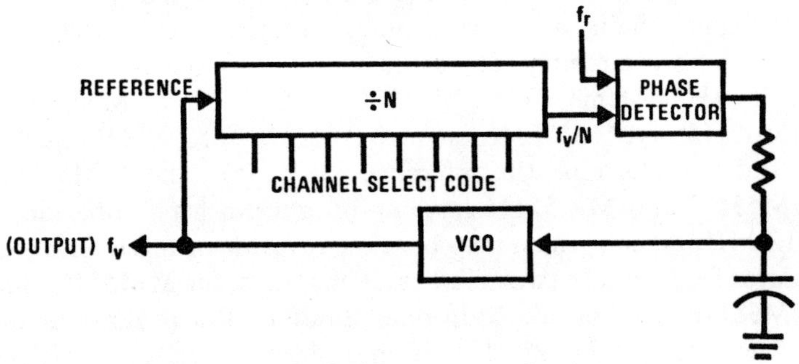

Fig. D-1. Basic frequency synthesizer.

Although simple in concept, the circuit of Fig. D-1 has certain difficulties. In CB, we are synthesizing the following frequencies:

Ch 1	26.965 MHz
Ch 2	26.975 MHz
.	.
.	.
.	.
Ch 40	27.405 MHz

Although the channel spacing is 10 kHz, a reference frequency of 5 kHz would be necessary due to the odd 5 kHz in the assigned channel. This in itself poses no problem; however, present technology limits the counting speed of programmable dividers to something less than 5 MHz, ruling out the approach shown in

Fig. D-1. Two solutions to this problem are shown in Figs. D-2 and D-3.

Frequency prescaling shown in Fig. D-2 reduces the VCO frequency by M (a fixed number) to a frequency that can be divided by the programmable counter. The reference frequency fr must also be reduced by M. In the case of CB, if M = 10, fv = 26.965 MHz, the input to the programmable divider will be 2.6965 MHz, and the 5-kHz reference frequency will be reduced to 500 Hz. This poses problems in speed of response of the phase locked loop.

The second technique mixes the output frequency of the VCO with a stable fixed frequency to obtain a related reference frequency.

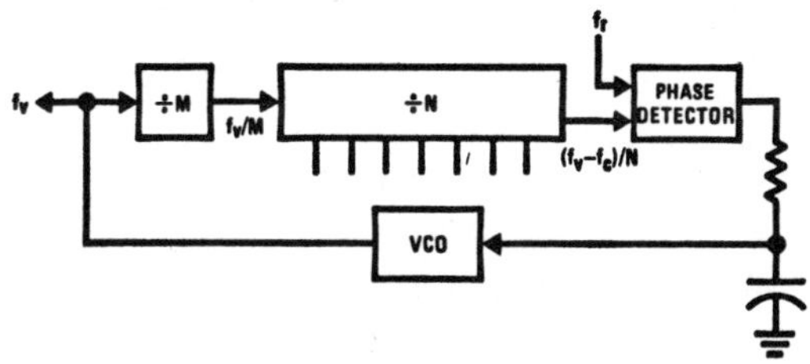

Fig. D-2. Frequency prescaling.

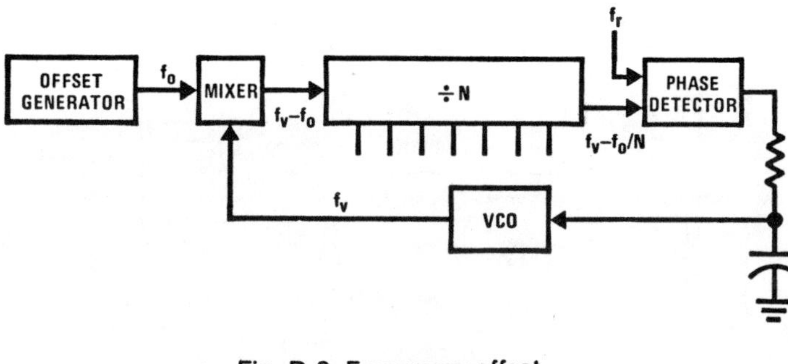

Fig. D-3. Frequency offset.

$$fv = Nfr + fo$$

This technique has the advantage of allowing a 10 kHz reference frequency in the loop instead of 5 kHz.

Further complexity arises when one considers that the synthesizer must also generate a local oscillator signal as well as a transmitter input signal for the radio (Fig. D-4). A system which

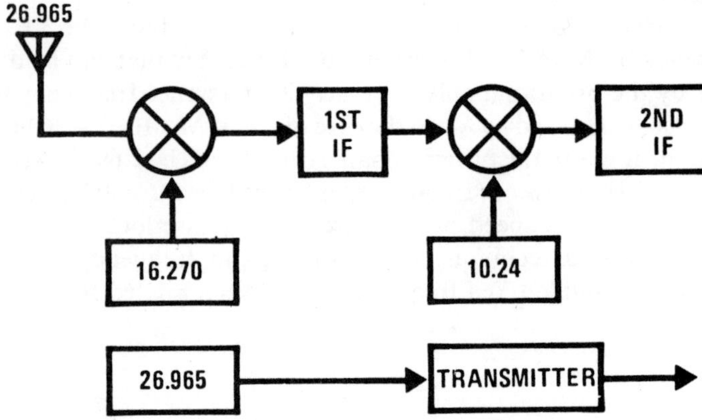

Fig. D-4. Signals needed to transmit and receive channel 1.

provides these frequencies, as well as the proper offset to allow the programmable divider to operate within its limits is shown in the typical applications diagrams (Figs. D-5 and D-6). The only de-

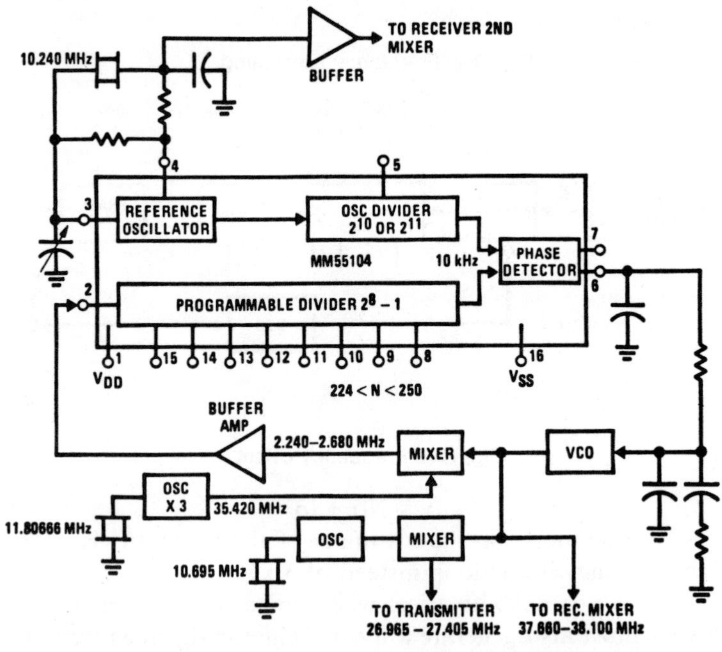

Fig. D-5. MM55104 or MM55114 three-crystal application.

parture from the ideal situation shown in Fig. D-4 is that the first i-f of 10.7 MHz must be changed to 10.695 MHz (a change of 5 kHz). See Figs. D-7 and D-8 for typical applications.

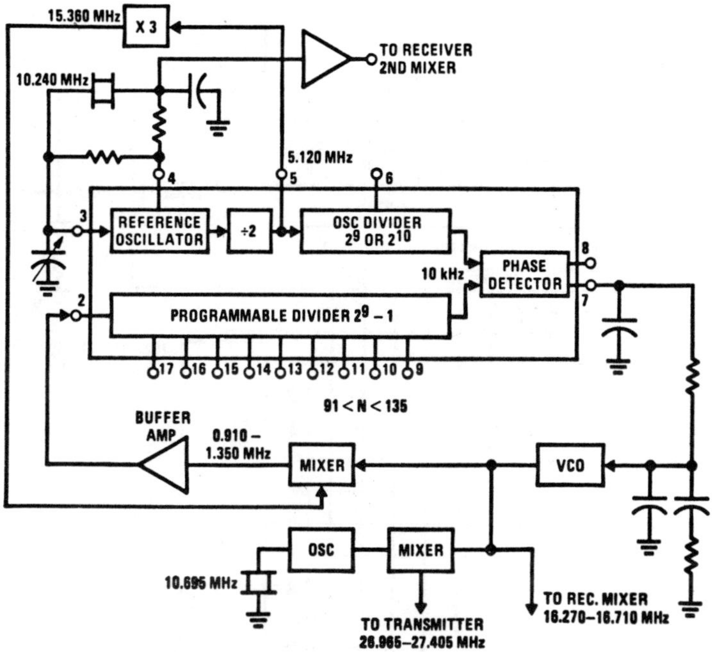

Fig. D-6. MM55106 or MM55116 two-crystal, 40-channel application.

MM55108, MM55110 PLL FREQUENCY SYNTHESIZER WITH RECEIVE/TRANSMIT MODE

The MM55108 and MM55110 PLL frequency synthesizers are monolithic metal gate CMOS integrated circuits which contain phase locked loop circuits useful for frequency synthesis applications in CB transceivers. The devices operate from a single power supply and contain an oscillator with feedback resistor, divider chain, a binary input programmable divider with control logic for the transmit mode (÷ by (N + 91)), and the necessary phase detector logic. The devices may be used in double i-f or single i-f systems.

Both the MM55108 and the MM55110 use a 10.24 MHz quartz crystal to determine the reference frequency. The MM55108 has a 2^{11} divider chain which generates a 5 kHz reference frequency. The MM55110 has a selectable 2^{10} or 2^{11} divider chain which gives either a 10 kHz or 5 kHz reference frequency.

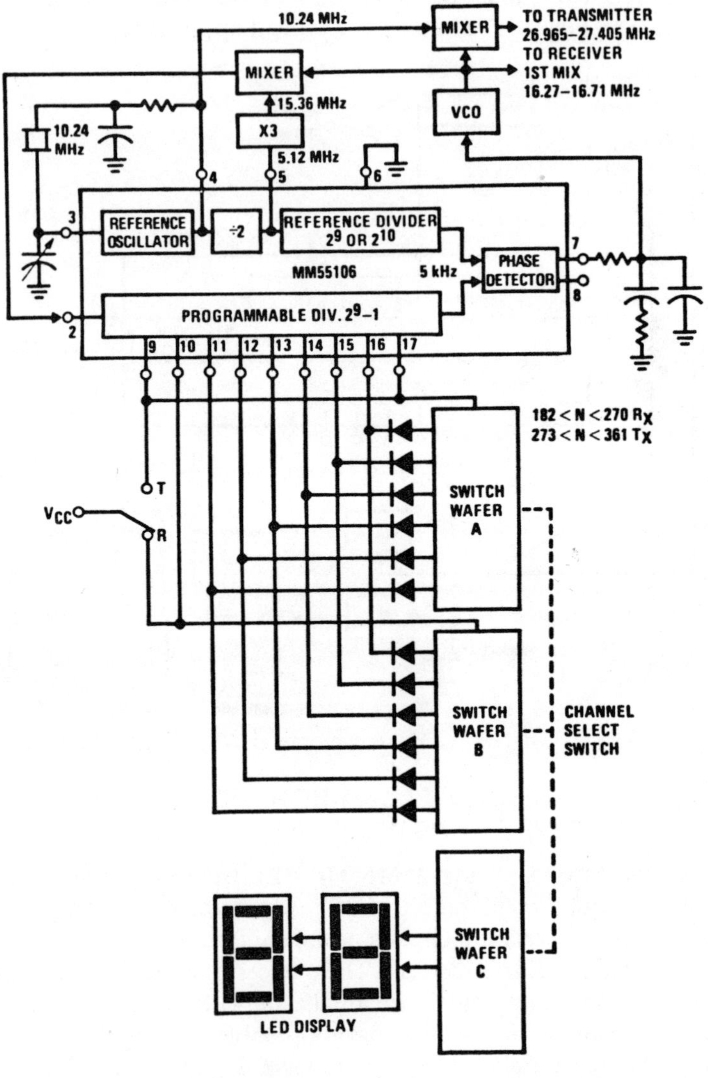

Fig. D-7. MM55106 or MM55116 single-crystal, 40-channel application.

The selection of reference frequency is made by use of the FS pin. In addition, the MM55110 contains an amplifier for filter applications and an additional input to the programmable divider which allows 2^{10} — 1 division of the input frequency (fIN) for FM applications. Due to the internal amplifier stage at input frequency

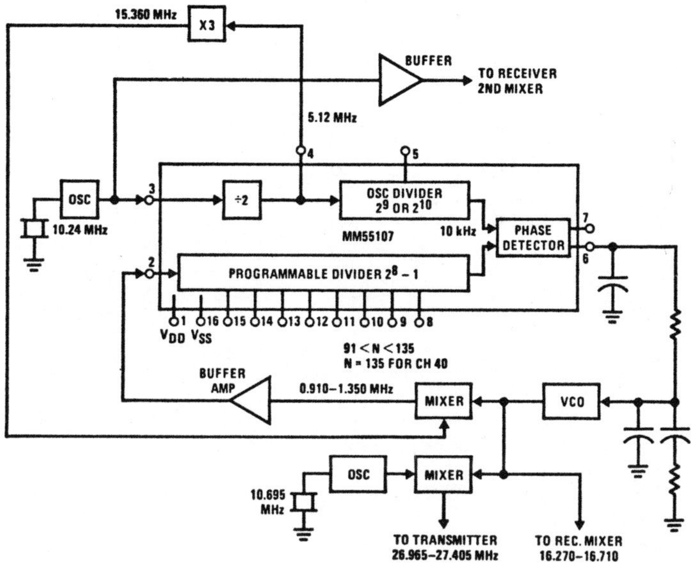

Fig. D-8. MM55107 two-crystal, 40-channel application.

input (fIN), the MM55108 and MM55110 may take a 0.5 Vp-p signal at fIN as the input frequency for the programmable divider. Inputs to the programmable divider are standard binary signals. Selection of a channel is accomplished by mechanical switches or by external electronic programming of the programmable divider. The ø VCO output provides a high level voltage (sources current) when the ø VCO frequency is lower than the lock frequency, and ø VCO provides a low level voltage (sinks current) when the ø VCO frequency is higher than the lock frequency. The ø VCO output goes to a high impedance state (TRI-STATE®) while in lock mode, and the lock detector output LD also goes to a high state under lock condition. Figures D-9 through D-11 show typical applictions.

Features

- ☐ Single crystal operation
- ☐ Single power supply
- ☐ Low power CMOS technology
- ☐ Binary input channel select code
- ☐ 2^{10} or 2^{11} divider chain from oscillator input (MM55110), 2^{11} divider chain (MM55108)

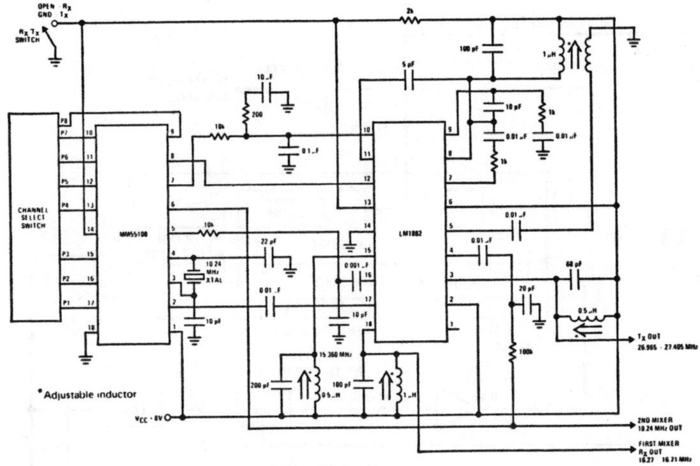

Fig. D-9. Single-crystal, 40-channel low side injection with MM55108 and LM1862.

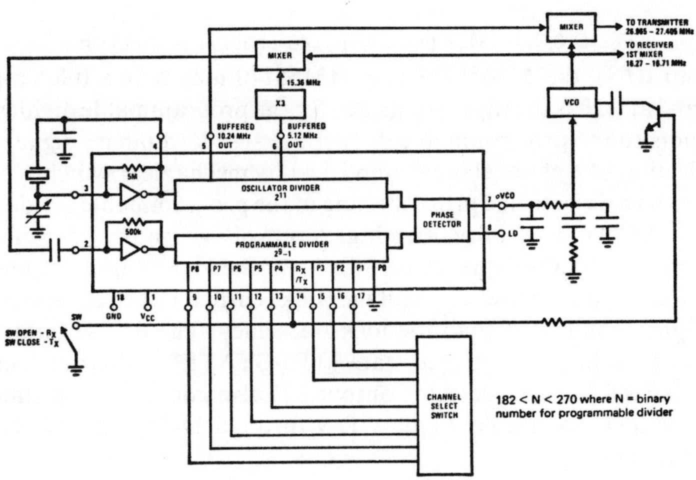

Fig. D-10. MM55108 single-crystal, 40-channel low side injection.

☐ Buffered 5.12 MHz and buffered 10.24 MHz outputs

☐ On-chip oscillator with bias resistor

☐ Pulldown resistors on programmable divider inputs

☐ Receive/transmit input for ÷ by (N+91) while in transmit mode

☐ Amplifier for filter applications (MM55110)

☐ Programmable $2^9 - 1$ division of fIN

☐ Additional programmable input for $2^{10} - 1$ division of fIN (MM55110)

☐ Amplifier stage on fIN input to accept 0.5 Vp-p signal

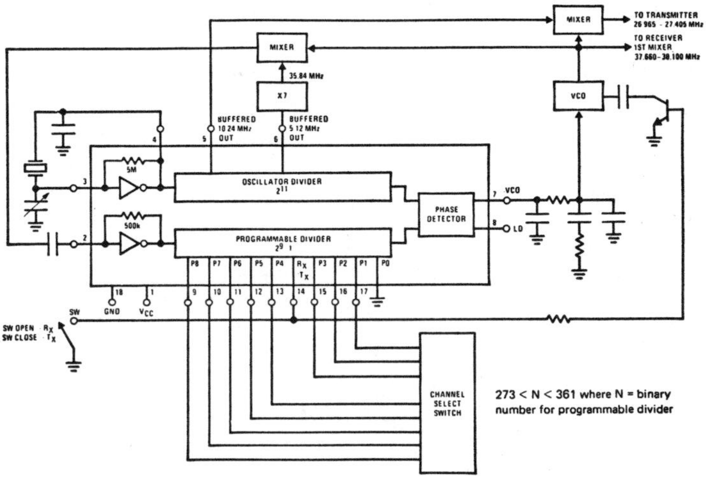

Fig. D-11. MM55108 single-crystal, 40-channel high side injection.

MM55109, MM55111 PLL FREQUENCY SYNTHESIZER AND CHANNEL PROGRAMMER

The MM55109 and MM55111 are phase locked loop frequency synthesizers utilizing metal-gate CMOS integrated circuit technology. These circuits are designed specifically for frequency synthesis applications in single crystal CB transceivers. A channel programmer (40 channels) with power-up reset to channel 19 and direct LED drive (MM55109) or direct vacuum fluorescent display drive (MM55111) are incorporated on-chip.

Channel selection is accomplished via the slew up/slew down feature with a fast/slow slew option (10 Hz/2 Hz) or via inputs from a quadrature switch.

The devices operate from a single power supply and contain a crystal oscillator stage with feedback resistor and a 2^{11} divider stage to generate the 5 kHz reference frequency from a 10.24 MHz quartz crystal. An amplifier stage is included at the frequency input (fIN) to accept a 0.5 Vp-p signal as the input frequency for the programmable divider. Input codes (N) for the programmable divider are generated on-chip for either the receive mode or transmit mode (N + 91) depending on the logic level at the receive/transmit input (Rx/Tx). Phase detector logic is also incorporated to accomplish the frequency synthesis. The phase comparator output (øVCO) provides a high level voltage (sources current) when the external VCO frequency is lower than the lock frequency, and the øVCO output provides a low level voltage (sink current) when the external VCO frequency is higher than the lock frequency. The øVCO output goes to a high impedance state (TRI-STATE®) when in the lock mode, and the Lock Detector output (LD) also goes to a high level voltage state under lock conditions. Figures D-12 through D-14 show typical applications.

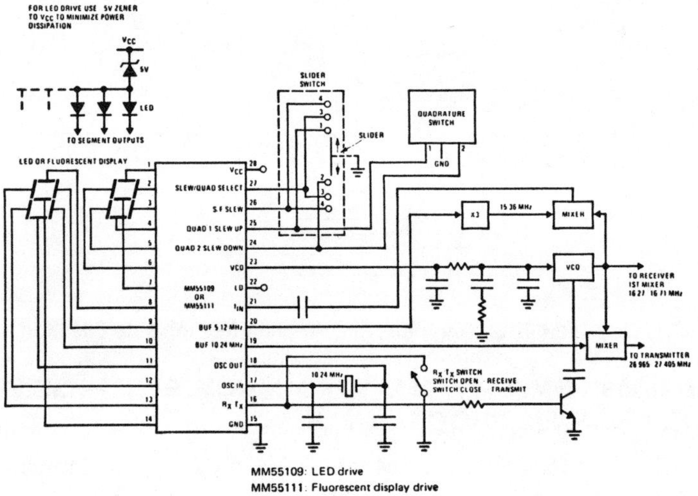

Fig. D-12. MM55109 or MM55111 single-crystal, 40-channel low side injection.

Features

- ☐ Single crystal operation
- ☐ Single power supply
- ☐ Low power CMOS technology
- ☐ 2^{11} divider chain from oscillator input
- ☐ On-chip oscillator with bias resistor
- ☐ Buffered 5.12 MHz output and buffered 10.24 MHz output
- ☐ Amplifier stage on fIN input to accept 0.5 Vp-p signal
- ☐ On-chip channel programmer for 40 channels
- ☐ Power-up reset to channel 19
- ☐ Receive/transmit input for ÷ by (N + 91) while in transmit mode

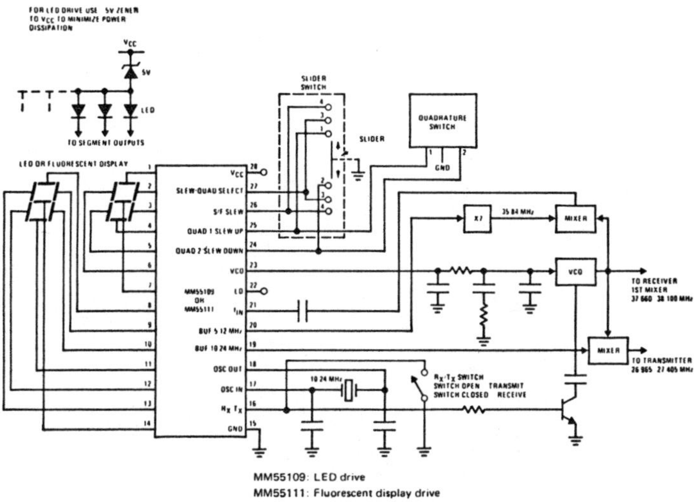

Fig. D-13. MM55109 or MM55111 single-crystal, 40-channel high side injection.

- ☐ Direct LED drive (ISINK) for MM55109
- ☐ Direct vacuum fluorescent display drive (ISOURCE) for MM55111
- ☐ Fast/slow slew (10 Hz or 2 Hz)
- ☐ Accept inputs from quadrature rotary switch for channel selection

MM55122 SERIAL DATA/PLL FREQUENCY SYNTHESIZER

The MM55122 is a monolithic metal gate CMOS integrated circuit which contains a phase-locked loop circuit useful for frequency synthesizer application in CB transceivers. The device operates from a single power supply and contains an oscillator, a 2^{10} divider chain, a binary input programmable divider, and phase detector circuitry. Selection of a channel is accomplished by external programming of the programmable divider with a 9-bit serial code derived from a 26-bit data string fed to the data 1/0 pin. The serial data format consists of a leading logical 1 synchronization bit, three 4-bit data to generate analog outputs (such as squelch, volume, or (avc), 4 control bits that are latched and made available at pins A-D, and a 9-bit binary input channel select code.

The MM55122 may be used in single or double 1F systems. It uses a 10.24 MHz quartz crystal to determine the reference frequency. It has an output pin which provides a 5.12 MHz signal,

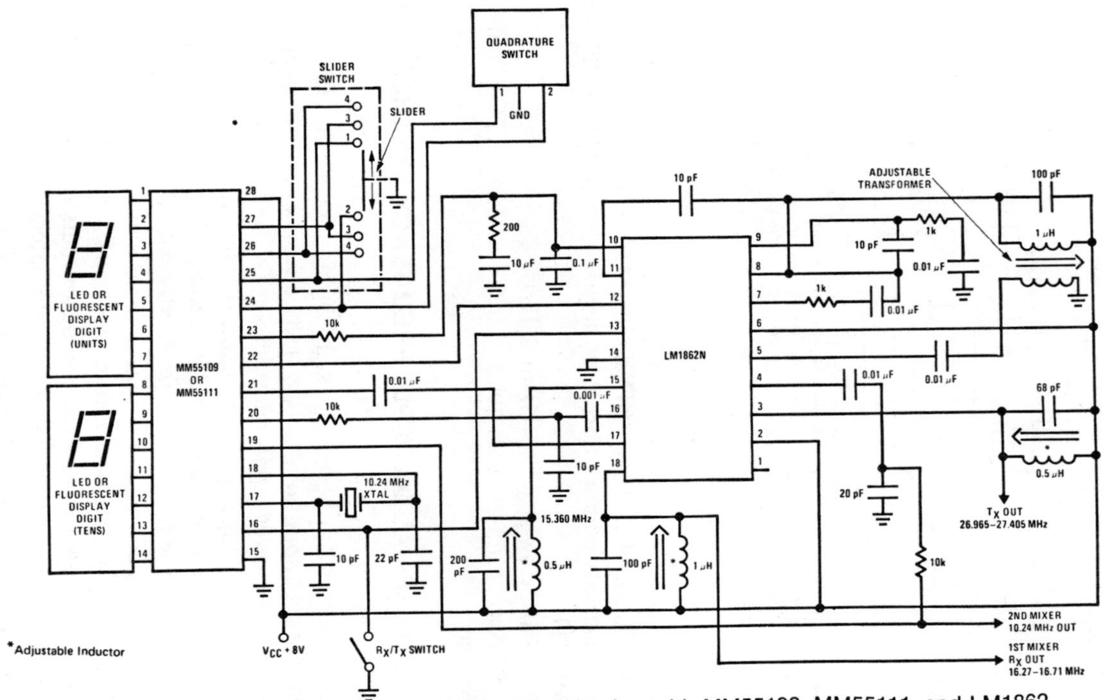

Fig. D-14. Single-crystal, 40-channel low side injection with MM55109, MM55111, and LM1862.

which may be tripled for use as a reference oscillator frequency in 2-crystal systems.

The phase detector output, øVCO, provides a high level voltage (sources current) when the VCO frequency is lower than the lock frequency, and it provides a low level voltage (sinks current) when the VCO frequency is higher than lock frequency. The øVCO output goes to a high impedance (TRI-STATE®) condition and the lock detector output, LD, goes to a high state under lock conditions.

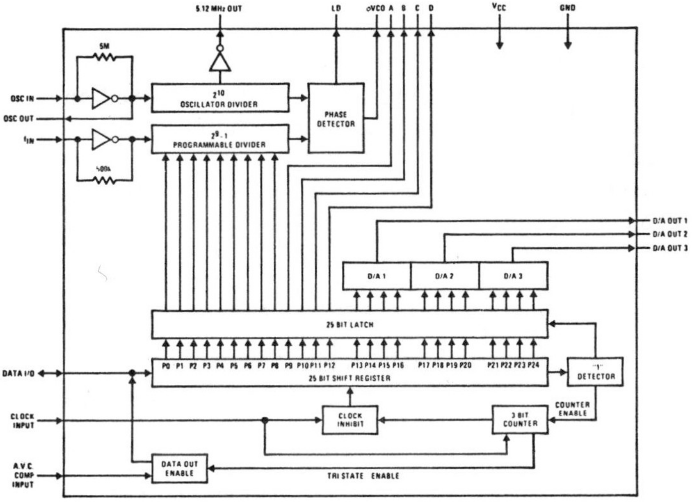

Fig. D-15. Block diagram of the MM55122.

Features

☐ Single crystal operation
☐ Single power supply
☐ Low power CMOS technology
☐ On-chip oscillator with feedback resistor
☐ Buffered 5.12 MHz output

Functional Description

The 10-kHz reference frequency of the phase detector is generated by division of a 10.24 MHz oscillator frequency in the 2^{10} divider chain. This is compared with the output of the programmable divider until a match in phase is reached to attain a lock condition. The inputs to the programmable divider are the VCO frequency (mixed down) input at fIN and the 9-bit binary channel selection code which is the divisor of the fIN input to generate a corresponding 10 kHz signal for the channel frequency in question.

Format of serial data generated in a controller and fed into the Data 1/0 pin is shown in Fig. D-16. From this data string, the logic 1 synchronization bit, three 4-bit data that are used to generate analog outputs, 4 control bits that are latched and made available at separate pins and 9-bit channel selection code are derived.

Synchronization is maintained between the controller-oriented processor and the MM55122 by use of a logical 1 hand-

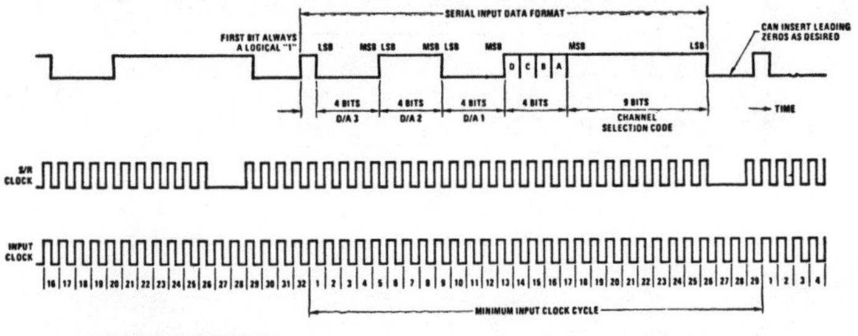

SYNCHRONIZATION STEPS:

CLOCK EDGE	OPERATION	CLOCK EDGE	OPERATION
26R*	Transfer data into latch	28R	Strobe Data I/O A.V.C. level

Fig. D-16. MM55122 operation.

S/R will be transferred (at the rising edge of the 26th clock pulse) to the 25-bit latch. The falling edge of the 26th clock pulse disables the transfer, enables the TRI-STATE data 1/0 for output mode, resets all of the S/R bits, and inhibits the clock internally to the shift register.

The rising edge of the 27th clock pulse disables the reset and will be used in the controller oriented processor to strobe the 1 data which will be present on the Data 1/0 pin.

The failing edge of the 27th clock pulse removes the 1 from the Data 1/0 pin and enables the avc comparator data. This data will be strobed into the controller by the rising edge of the 28th clock pulse. The failing edge of the 28th clock pulse disables the A.V.C. data output. The rising edge of the 29th clock pulse enables the data 1/0 as an input and the next clock pulse clocks data into the S/R. Typical applications are shown in Figs. D-17 and D-18.

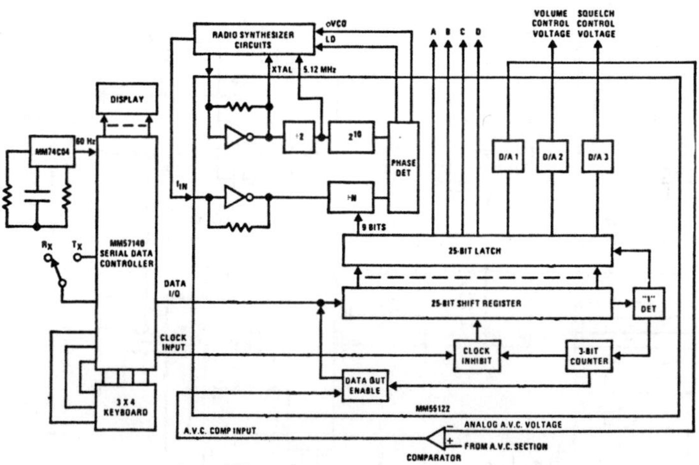

Fig. D-17. Remote microphone serial data system.

MM57150 STANDARD CB RADIO CONTROLLER

The MM57150 Standard Citizen Band (CB) Radio Controller was developed using a metal gate P-channel enhancement and depletion-mode MOS/LSI technology with a primary objective of low end-product cost. The MM57150 contains the logic necessary to scan a low-cost keyboard or momentary switches, provide control code inputs to a binary input PLL frequency synthesizer

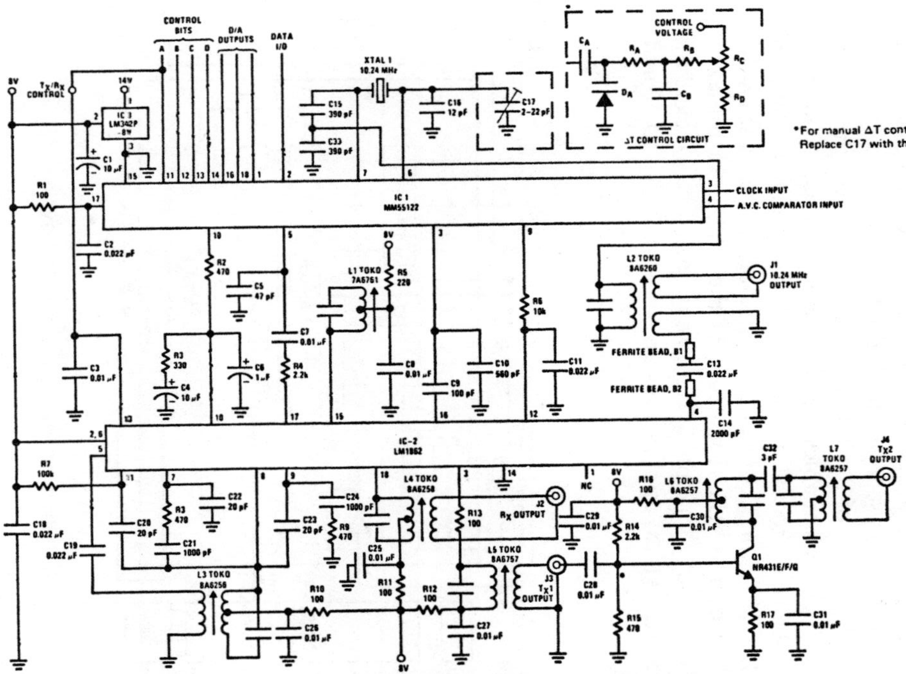

All resistors shown in ohms, ±5% tolerance. Capacitors in μF and pF, 80%–20% tolerance for μF, ±10% for pF.
Tuning capacitors parallel all transformers are built-in type with the coils. Details referred to Toko published specifications.

Fig. D-18. LM1862, MM55122 single-crystal, 40-channel PLL.

(i.e., MM55116 or equivalent) for 40-channel operation and provide display data for a 2-digit channel number display. Figure D-19 shows the interconnect for the MM57150, keyboard switches, display and PLL frequency synthesizer.

The MM57150 operates from a single power supply and directly drives the MM55116 and CD4511 display drivers. The device will scan a 4 × 4 keyboard matrix, key closures in the form of slide switches, rocker switches, momentary close push buttons, or standard keyboards, inputs from which will step the controller up or down, fast or slow through all 40 channels. Additionally, up to 10 special channels can be loaded into user programmable memories. In monitor mode, one user selectable channel can be scanned on a regular basis while monitoring another channel. Single key operation allows the user to immediately switch to emergency channel 9. During receive, the MM57150 provides the proper output channel codes to the MM55116 for generation of the 16.27 to 16.710 MHz (low side) or 37.66 to 38.1 MHz (high side) input to the first mixer for receiving, or the codes necessary to generate the 26.965 to 27.405 MHz output to the transmitter during transmit. During transmit and receive, the display outputs present BCD data which is latched in the CD4511 display drivers to display the 2-digit channel number.

In memory operation, a 1-to-10 wrap-around counter is used to index the active memory. Depressions of the "MEM" key will advance the memory index, 1, 2,...9, 10, 1.

Features

- ☐ Low power MOS technology, approximately 8 mA standby
- ☐ Slow 4-channel per second slew up or slew down
- ☐ Fast 8-channel per second slew up or slew down
- ☐ Full 40-channel capability
- ☐ Favorite channel background monitoring capability
- ☐ 10 user programmable memory channels
- ☐ Scan all/scan memory mode
- ☐ Direct keyboard channel selection
- ☐ Single key depression for emergency channel 9 selection
- ☐ High or low side conversion for RECEIVE
- ☐ Squelch controlled option
- ☐ Outputs limited to 40 legal channels
- ☐ Lock-on channel during TRANSMIT
- ☐ Initialized to channel 19 on power-on
- ☐ Direct key entry for alternate channel

Key Closure Description

The HELP key terminates any previous mode, scan, monitor, etc. and sets the programmer outputs and display to channel 9.

Depression of the ALL stops scan, and monitor, setting the controller in the 40-channel mode. The active channel is switched to the last channel active in the "ALL" mode.

MEM sets the controller to the Memory mode, terminating any scan or monitor mode. If non-zero. the previously active memory becomes the active channel. Subsequent depressions advance the memory index, changing the active channel to the channel stored in the next memory. The display will read 00 if there is no channel stored in a given memory.

Depressing ALT halts any scanning or monitoring and resets the memory mode. The alternate channel and active channel are then interchanged if the active channel is non-zero. The ALT key allows the user to select the channel to be used for the monitor mode.

Depression of scan while in the ALL 40 mode initiates a sequential wrap-around scan of channels 1—40 beginning with the present active channel. The scan direction (up or down) will be identical to the direction of the last slew key depressed. If the controller is already scanning when the SCAN key is depressed, the scan will stop.

While scanning, the controller increments the active channel and waits 250 ms on the new channel before testing the SQ input. If the SQ input is logic high (VCC — signaling no activity on this channel), the scan will continue. If the SQ input is logic low (ground — signaling channel activity) the scan mode will be reset and scanning will not resume until the SCAN key is depressed again.

If SCAN is depressed while in the memory mode, a scan of the 10 memory channel will begin. If a memory channel is empty, (00), the controller will jump to the next valid memory channel and continue scanning.

If not already in the monitor mode, MON will place the controller in this mode. The alternate (ALT) channel will be tested after 250 ms, and then again every 10 seconds. If the SQ input is low while testing the alternate channel, the alternate channel will become the active channel until SQ goes high. When SQ does go high, the original active channel will be restored and the monitor mode resumed. Any key depression during the 250 ms wait to test

SQ will also restore the active channel. If the XMIT input goes low while testing the alternate channel, the monitor mode will be terminated. Note that the alternate channel is now the active channel, and the previous active channel is now stored in the ALT memory.

A depression of the MON key while the monitor is already engaged will reset the monitor mode. Also, the controller will not enter the monitor mode if the active channel is 00. In this case, the alternate channel will replace the active channel.

In the non-memory mode," ▲"," ▼" scroll the active channel up or down 1 channel. If the key remains depressed for more than 500 ms, the active channel will scroll up or down at the rate of 4 channels per second. In the memory mode, the active channel will also be loaded into memory, making the scroll 01, 00, 40. This allows the use of 00 to clear a memory.

The keys, "⇧", "⇩" scroll the active channel up or down at a rate of eight channels per second.

In the non-memory mode, CH sets the entry mode. The first digit entry clears the active channel number from the display, displaying the number entered. Subsequent number entries shift the display from right to left (LSD to MSD). After 2 seconds of keyboard inactivity, the entry mode will then be reset. The validity of this entered number is then tested, and if valid, becomes the active channel. If the entered number is not valid, the active channel will be set to channel 19. If the user does not wish to wait 2 seconds, a second depression of the CH key will reset the entry mode, and immediately initiate a validity test on the entered channel number. A low on the XMIT input will also reset the entry mode.

In the memory mode, a valid entry will become the active channel and will be entered in the active memory. 00 can be entered. In entry mode, the displayed entry does not reflect the channel code to the PLL until the entry mode is reset. Depression of any non-number key while in the entry mode will reset any previous activity.

Logic Inputs

K1-K4 are used to sense key presses on the keyboard. These inputs are debounced for approximately 10 ms on both contact and release.

The HI SIDE input is used to select between 2 sets of channel codes presented to the PLL. A logic low on HI SIDE will output the Low Side Channel Code on P0—P8, while a logic high will

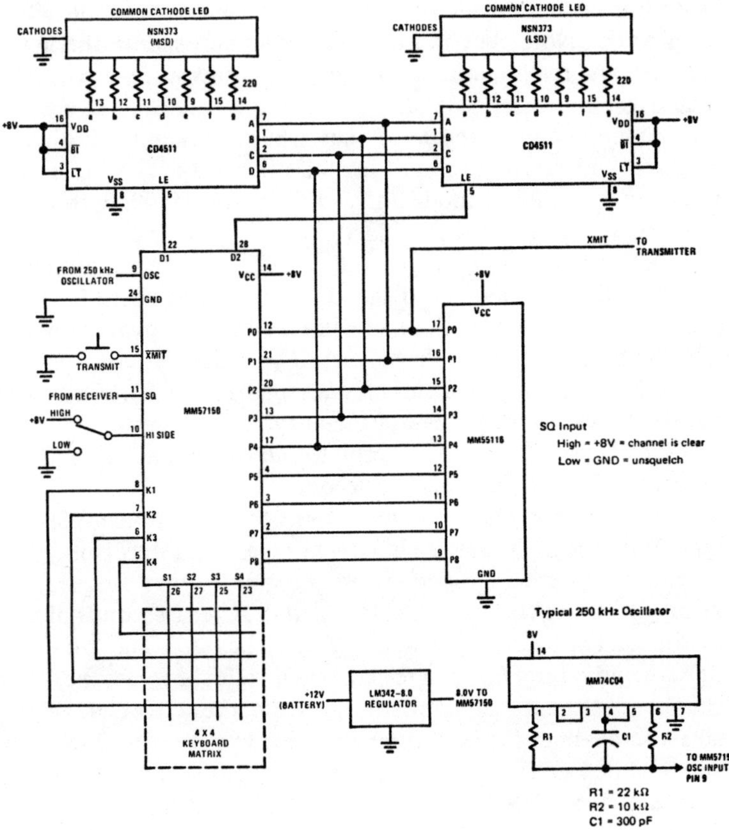

Fig. D-19. CB radio controller with 40 channels.

force the controller to output the High Side Channel Code on P0—P8. A logic transition on HI SIDE is only reflected by the channel code after a change of the active channel or a low on the XMIT input.

The SQ signal is generated by the CB receiver and is used by the MM57150 to determine if there is a signal present on the active channel. The status of the SQ input is only relevant during the scan or monitor modes. As stated previously, a low on SQ will terminate a scan but only delay a monitor.

The XMIT input is used to place the controller in either the transmit or receive modes. A logic low (ground) on the XMIT input will place the controller in the transmit mode, terminating any other modes. The channel code (see Table 1), represented by outputs P0—P8, will then change to a binary value associated with the transmission frequency of the active channel. The controller will not respond to any other input or key closure as long as XMIT is low. A logic high (VCC) on XMIT greater than 1 ms duration will terminate the transmit mode and restore the receive channel code on outputs P0—P8.

Logic Outputs

D1 and D2 are used for enabling the latches in the left and right channel display drivers. Normally high, D1 and D2 go low for a short time interval after BCD data has been presented to the drivers (CD4511 or equivalent) via outputs P1-P4 (Fig. D-19).

P0-P8 are used for presenting the 9-bit channel code to the PLL (see Table 1). As aforementioned, P1-P4 are also used for supplying BCD data to the display latches. The LSB of the channel code, P0, gives high only when the controller is in the transmit mode (XMIT = 0).

The four outputs are used in conjunction with K1-K4 to service 16 keypads. A key press is achieved by decreasing the resistance between any K and S line to less than 5 K Ω for a time period exceeding 10 ms. The keyboard matrix configuration section specifies which keyboard function is associated with each intersection of the S outputs and K inputs.

Proper operation of the MM57150 requires that the power supply turn-on time be less than 1 ms. In applications where the available power supply has a very slow turn-on time. The circuit shown below can be used to buffer the MM57150, MM55116 supply.

Appendix E Fairchild 11C84 PLL IC

The following data is reprinted by permission of Fairchild Camera & Instrument Corp.

The 11C84 is a 16-pin LSI chip containing all the digital circuitry normally used in a digital PLL, including a programmable divide-by-N counter, a reference frequency divider chain, a phase/frequency comparator and a 10.24-MHz oscillator circuit. It accepts channel select inputs from BCD coded switches or the electronic equivalent. Also provided are an out-of-lock output signal and a 5-kHz offset control for single-sideband applications.

External circuitry required to complete a PLL are a voltage controlled oscillator (VCO), an active integrator and a 10.24-MHz crystal. The 11C84 is made with low power Schottky technology to achieve high frequency dividers. This eliminates any need for a prescaler or down-converter between the VCO output (16 to 17 MHz) and the divide-by-N counter input.

The 11C84 is designed for use in dual-conversion (10.695 MHz first i-f, 455-kHz second i-f) receivers. The transmit frequency is derived by mixing, rather than being generated directly by the VCO, thus avoiding the problem of feedback from the modulated rf power amplifier back into the VCO. The IC offers:

- ☐ BCD Channel Selection
- ☐ Pull-up resistors on channel select inputs
- ☐ Works with single or dual crystal systems
- ☐ PLL out-of-lock output available, to inhibit transmission
- ☐ Designed for dual conversion receivers
- ☐ Low power Schottky technology-operates directly at VCO Frequencies
- ☐ Self-biased divider input simplifies capacitive coupling from VCO
- ☐ +5 kHz offset control for SSB applications
- ☐ Transmit frequency derived by mixing - eliminates feedback-induced distortion
- ☐ Operates from +5 VDC supply
- ☐ 16-pin DIP package

INPUT AND OUTPUT FUNCTIONS

There are seven channel select inputs intended for use with a

2-digit BCD switch or the electronic equivalent. The four inputs U_0-U_3 accept the units BCD numbers 0-9, while the three T_0-T_2 inputs accept the tens BCD numbers 0-7. Each of these seven inputs has a 20K pullup resistor to V_{cc}, for use with a switch that provides contact closures to ground.

The mode control input M is provided for optional use in single sideband (SSB) applications. Grounding the M input increases the VCO frequency by 5 kHz in both transmit and receive modes, for SSB operation. For normal (non-SSB) operation, M should be left open; an internal pull-up resistor establishes a HIGH signal level when M is open.

Terminal VF is one input of a differential amplifier whose other input is connected to an internal reference voltage (Fig. E-1). The VF input resistance is typically 8K Ω and is biased in the active region to facilitate capacitive coupling from the VCO. It accepts either sinusoidal or rectangular waveshapes and responds to peak-to-peak amplitudes of 500 mV.

The transmit receive (TR) input changes the divide ratio of the programmable divider, to coordinate with the 455 kHz change in VCO frequency when changing between the two modes. The T/R input should be grounded for the transmit mode and open for the receive mode. The T/R input contains a 20K Ω pullup resistor to establish the locgic HIGH level in the open condition.

The 11C84 uses a 5-kHz comparison frequency, obtained by digitally dividing down from a 10.24 MHz reference frequency. The 10.24-MHz signal may be generated externally or internally. If the transceiver uses a separate 10.24-MHz oscillator, its output should be connected to terminal X_1 of the 11C84 and can be capacitively coupled. A peak-to-peak amplitude of 500 mV (Typ) is sufficient. The 10.24-MHz signal can be generated internally by connecting a crystal across terminals X_0 and X_1, as shown in Fig. E-1.

The channel select, M and T/R inputs, in addition to having pull-up resistors to V_{cc}, are low-power Schottky TTL-type inputs, as shown in Fig. E-1. The inverters connected to X_0 and X_1 are also of this type. The quiescent HIGH and LOW logic levels applied to the Channel Select, M and T/R inputs should meet the same criteria specified for TTL logic; a HIGH signal should be 2.4 V or more, while a LOW signal should be 0.4 V or less. These values guarantee adequate noise margins. An input should not be driven more positive than + 10 V.

The detector output (DO), as shown in Fig. E-1, has three

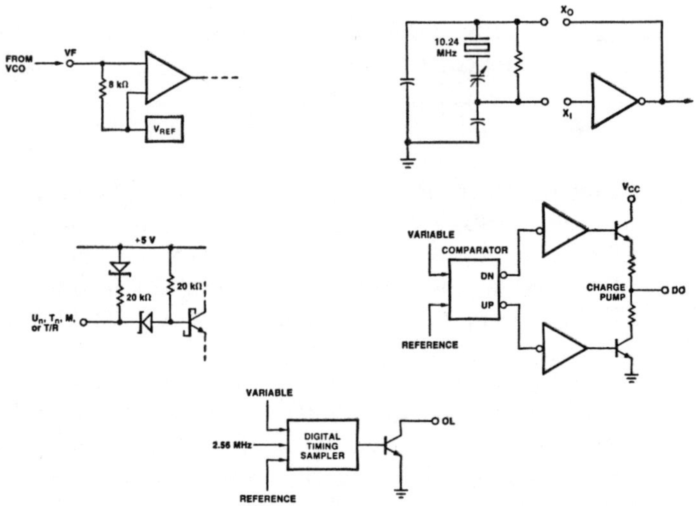

Fig. E-1. Input and output configurations.

possible conditions: current sourcing, current sinking and high impedance OFF state. When the loop is locked, DO is in the OFF state and generates no error signals to the integrator. When the variable frequency input to the phase/frequency comparator differs from the 5 kHz reference, the DO output is a series of current pulses with a 5 kHz repetition rate and a duration equal to the phase error. The DO output pulses are negative-going, i.e., current sinking, when the variable frequency lags the reference frequency and are positive-going when the variable frequency leads the reference frequency. These DO output pulse polarities insure the proper direction of varactor-bias change after inversion in the integrator. The phase/frequency comparator cannot give a false indication of lock when one input frequency is a multiple of the other. The comparator also has an anti-backlash feature that prevents it from operating in the dead zone. For this reason, the charge-pump output will always exhibit a short negative-going pulse followed immediately by a short positive-going pulse. the duration and proximity of these opposing pulses cause no net change to the charge on the integrator.

The out-of-lock output is an open collector transistor (Fig. E-1), which is turned off in the locked condition. The transistor is turned on whenever the negative going edges of the comparator input signals differ by ≥ ± 200 ns or more. The transistor will then stay on until such time as these negative edges occur within 200 ns of each other for at least two cycles of the 5 kHz reference frequency. The ± 200 ns tolerance allows for leakages associated with the integrator and ensures that the integrator only responds to true out-of-lock conditions. Typically, the OL output is used to suppress transmission and provide an indication of malfunction.

SINGLE-CRYSTAL SYSTEM

The VCO frequency is injected into first mixer of the receiver and is 10.695 MHz (the first intermediate frequency) below the selected channel frequency. The 10.24-MHz crystal frequency is injected into the second mixer to obtain the 4550kHz second i-f. In the transmit mode, the VCO output is mixed with the 10.24 MHz to obtain the selected channel frequency. See Fig. E-2.

In a single-crystal system, the VCO must cover both the receive and the transmit tuning ranges, which are offset by 455 kHz. The total required tuning range is approximately 1.5 MHz. To achieve this broad range without incurring penalties in lock-up time when changing between Receive and Transmit modes, an extra varactor diode is effectively switched in or out, respectively. The bias voltage on the extra varactor is designed (or adjusted) so that the VCO control voltage does not change when switching between modes, thus reducing lock-up time.

DUAL-CRYSTAL SYSTEM

In the receive mode, operation is the same as in the single crystal system. In the transmit mode, a second crystal oscillator (10.695 MHz) is mixed with the VCO frequency to derive the selected channel frequency. This means that the VCO frequency is the same in either the Receive or the Transmit mode, for a given channel. Thus no band-switching is required; the tuning range need be only about 500 kHz and higher Q tuned circuits can be used to achieve a cleaner VCO design. Also, longer time constants can be used in the loop integrator, thus suppressing the 5 kHz (reference frequency) sideband components in the VCO output. See Fig. E-3.

COST VERSUS PERFORMANCE TRADE-OFF

Although the dual crystal approach requires, of course, a crystal and the associated oscillator components, there is actually very little difference in parts count because the single crystal

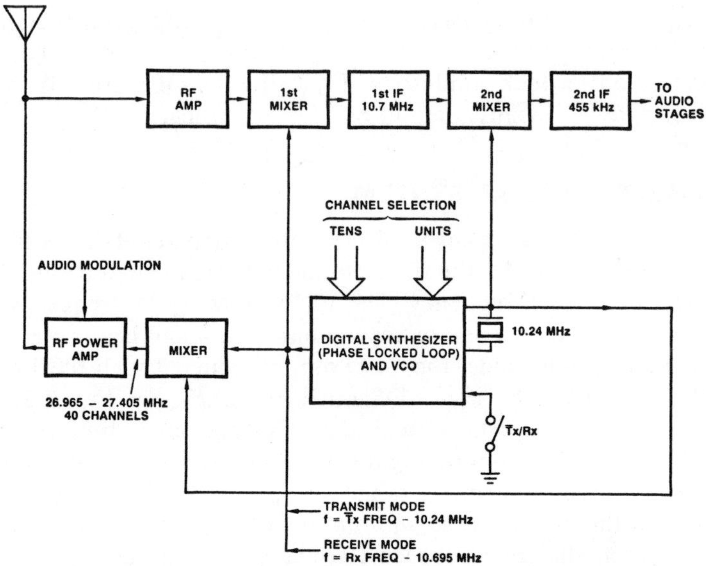

Fig. E-2. Single-crystal control.

system requires an extra varactor, potentiometer, switching transistor and auxiliary components. The disadvantages of increased complexity and lower performance of a band-switched VCO tends to dissipate the slight economic advantage of the single crystal system.

40-CHANNEL CB SYNTHESIZER

Figure 4 shows a synthesizer using the single crystal approach and a derived transmit frequency. This circuit was developed while the 11C84 was still in design, using MSI logic functions to simulate the 11C84. In this simulation it was found that digital noise from the programmable divider, feeding back into the VCO, caused spurious noise problems in the receiver. Buffer stage Q9 was incorporated to prevent this problem. Similarly, buffer stage Q6 isolates the 10.24-MHz oscillator from the reference frequency divider input. In using the 11C84, rather than its simulation, these buffer stages may or may not be required.

The primary objective of the design was to provide spectrally clean VCO and 10.24-MHz frequencies. To accomplish this, it

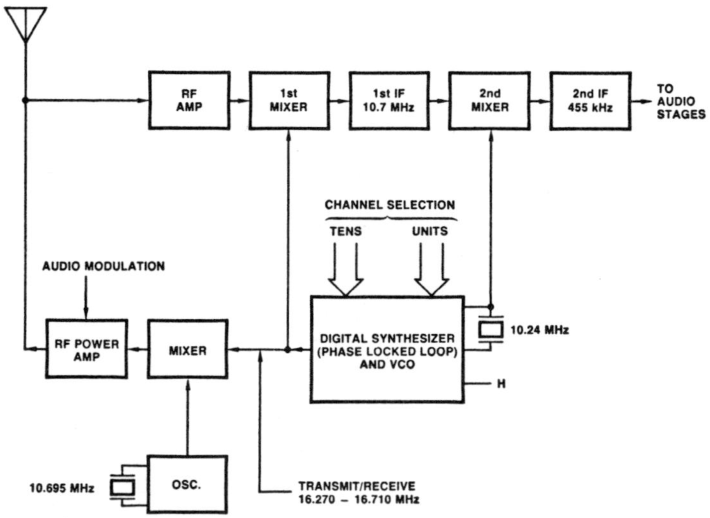

Fig. E-3. Dual-crystal control.

was necessary to separate the grounds for the digital and analog portions of the circuit. Therefore, rf transformers are used between the oscillators and the buffers. Transformer coupling also makes it possible to physically separate and shield the 11C84 and buffers to reduce digital noise radiation. The remaining circuitry can, of course, be located on the main pc board of the transceiver. Decoupling and grounding techniques are of the utmost importance.

TRANSMIT FREQUENCY GENERATION, DIRECT VERSUS DERIVED

As in all system design, trade-offs must be made in the various performance characteristics. Since the 11C84 input circuitry and counters are designed to operate at frequencies up to 30 MHz, a system could be built in which the VCO operates directly at 27 MHz. But, it is very difficult to prevent feedback to the audio-modulated rf-transmitter output into the VCO which, in turn, causes spurious FM modulation of the VCO. For this reason, the 11C84 was programmed for a system using mixing techniques to

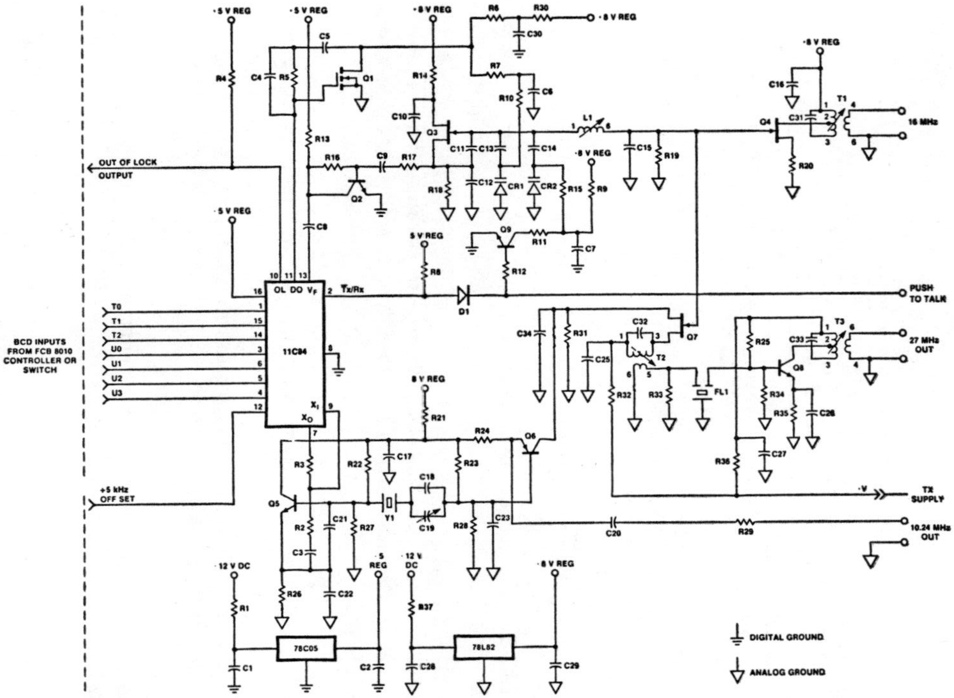

Fig. E-4. Synthesizer that uses the single-crystal approach and a derived transmit frequency.

derive the transmit frequency. That is, the VCO operates at approximately 17 MHz and is mixed with 10.24 MHz to obtain the 27-MHz transmit frequency. This generates a clean VCO signal, free of FM-modulation, but introduces sum and difference mixing products, called spurs, that require filtering.

One advantage of using mixing to derive the transmit frequency is that the VCO need change by only 455 kHz between transmit and receive; this can easily be accomplished by switching in an extra varactor diode. In a system where the transmit frequency is derived directly, the VCO must change by 10.24 MHz between Transmit and Receive thereby requiring complex inductive switching. See Fig. E-4.

Index

A

Active filter 193
AM receive mode 144
  transmit mode 146
Automatic frequency control 65
  phase control switch 65

B

Bandpass filter 138
BCD 9
Bias gate 236
Binary coded decimal system 9

C

Channel 40 operation,
  circuit analysis 96, 118

Charge pumps 14
Clarifier control 138
Cobra 1000 GTL CB synthesizer 32
Code converter 12, 158
Common cathode 62
Control voltage 54
Correction voltage 54
Crystal-controlled oscillators 16

D

Delta-tune 207
  control 112
Detector, phase 103
Digital synthesizer 92
Dividers, fixed 13
  programmable 7, 103, 231
  programming 10
  reference 103
Down mixer 211
  oscillator 113, 210, 236

E

E.F. Johnson Messenger
  4140 synthesizer 48

F

Fanfare 190 DF PLL synthesizer 67
Filters 14
Fixed dividers 13
Frequency control, automatic 65
Frequency/phase detector 43
Frequency sythesis 250

I

Input mode switch 158

L

Least significant bit 8
LED channel display 117
  circuit 144, 194, 212
Lock detectors 14, 66
Logic inputs 267
  outputs 268
Loop driver 172
  mixer 172, 192
  oscillator 192
  oscillator/Delta tune 171
Lower sideband receive mode 148
  transmit mode 149
LSB 8

M

Maximum divisor 19
Messenger 4140
  LED digital readout 62
Midland 77-839
  CB PLL synthesizer 86
Minimum divisor 19
Mixers 16
MM55104N, IC-1 108

practical application 104
MM55106, internal structure 178
practical application 182
special features 178
MM55122 Serial Data/PLL
frequency synthesizer 259
MM57159 Standard
CB Radio Controller 263
MN6040, internal structure 203
special features 201
Modular construction 74
Most significant bit 8
MSB 8
muPD858C, internal structure 123
practical application 127
special features 123

O

Offset buffer 116
crystal oscillators 138
mixer 20, 115
Oscillators, crystal-controlled 16
down 113, 210, 236
loop 192
offset 20, 113
reference 103, 167, 207
transmit 173, 211
voltage-controlled 14, 167
Out-of-lock detector 142

P

Pearce-Simpson Super Tiger
40A PLL synthesizer 107
Phase comparator 235
control switch, automatic 65
detectors 14, 20, 65, 103
PLL amplifier 142
as a system 17
considerations 21
IC 167, 190
mixer 142, 237
synthesizer IC 207
troubleshooting 36, 57, 76, 99, 120, 152, 176, 198, 216, 238
PLL02A, internal structure 79
practical application 82
special features 79
President Model Madison 134
Programmable
dividers 7, 20, 65, 103, 231
Programming inputs 43
Pure binary system 8

R

Realistic TRC-424 166
REC86345, internal structure 63
special features 63
Reference divider 103
oscillator 103, 167, 207
oscillator stage 86
section 63
Regency CB-501 PLL circuit 190
Regulated 5-volt supply 96
Robyn SX-402D
PLL synthesizer 207

S

SC42502P/3001-201,
internal structure 43
special features 42
typical application 46
Sharp Model
CB-2260 PLL synthesizer 231
Stabilizer circuit 173

T

TC5080P 218
TC5081P 222
TC5082P 225
10.240-MHz buffer 192
Tram D12 PLL synthesizer 225
Transmit inhibit 211
mixer 173, 193, 211, 237
oscillator 173, 211, 237
stop circuit 89, 142
Transmit frequency 138
lock stage 113
Transmit killer 193
unlock detector 172
Transmit oscillator/mixer 89
IC-3 116

U

Unlock detector 211
uPD2816C, internal structure 24
special features 24
typical application 29
uPD858C, IC PLL stage 134
uPD861C, internal structure 158
practical application 164
special features 157
Upper sideband receive mode 146
transmit mode 147

V

Varactor diode 14
VCO 14, 235
buffers 210, 235
& mixer 86
stage 112, 135
Voltage-controlled
oscillator 14, 167
& buffer 191